D1279647

The Study of Educational Politics

Education Policy Perspectives

General Editor: Professor Ivor Goodson, Faculty of Education, University of Western Ontario, London, Canada N6G 1G7

Education policy analysis has long been a neglected area in the UK and, to an extent, in the USA and Australia. The result has been a profound gap between the study of education and the formulation of education policy. For practitioners, such a lack of analysis of new policy initiatives has worrying implications, particularly at a time of such policy flux and change. Education policy has, in recent years, been a matter for intense political debate – the political and public interest in the working of the system has come at the same time as the breaking of the consensus on education policy by the New Right. As never before, political parties and pressure groups differ in their articulated policies and prescriptions for the education sector. Critical thinking about these developments is clearly imperative.

All those working within the system also need information on policy-making, policy implementation and effective day-to-day operation. Pressure on schools from government, education authorities and parents has generated an enormous need for knowledge among those on the receiving end of educational policies.

This Falmer Press series aims to fill the academic gap, to reflect the politicalization of education, and to provide the practitioners with the analysis for informed implementation of policies that they will need. It offers studies in broad areas of policy studies, with a particular focus on the following areas: school organization and improvement; critical social analysis; policy studies and evaluation; and education and training.

The Study of Educational Politics

The 1994 Commemorative Yearbook
of the
Politics of Education Association
(1969–1994)

Edited by

Jay D. Scribner
The University of Texas at Austin

and

Donald H. Layton
The University at Albany
State University of New York

 The Falmer Press

(A member of the Taylor & Francis Group)
Washington, DC • London

UK   The Falmer Press, 4 John Street, London WC1N 2ET
USA   The Falmer Press, Taylor & Francis Inc., 1900 Frost Road, Suite
        101, Bristol, PA 19007

First published 1995

A catalogue record for this book is available from the British Library

Library of Congress Cataloging-in-Publication Data are available on
request

ISBN 0 7507 0418 7 cased
ISBN 0 7507 0419 5 paper

Jacket design by Caroline Archer

Typeset by RGM Typesetting, The Mews, Birkdale Village,
Southport PR8 4AS, England

*Printed in Great Britain by Burgess Science Press, Basingstoke on paper which
has a specified pH value on final paper manufacture of not less than 7.5 and is
therefore 'acid free'.*

Contents

Politics of Education Association
Yearbook Sponsor

The **Politics of Education Association (PEA)** promotes the development and dissemination of research and debate on educational policy and politics. PEA brings together scholars, practitioners, and policy makers interested in educational governance and politics. It is affiliated as a Special Interest Group with the American Educational Research Association (AERA), and meets each spring in conjunction with AERA's annual meeting. PEA also sponsors its own biennial conferences on current policy issues in education in the fall. The annual membership dues for PEA are $US25.00 (subject to change). Members receive a copy of the annual *Yearbook* and the *Politics of Education Bulletin*, which includes news on member activities and occasional short scholarly pieces. Membership dues should be sent to **Louise Adler, PEA Treasurer, EC 552, Educational Administration, California State University, Fullerton; Fullerton, CA 92634–8000, USA**. Previous PEA *Yearbooks* and their editors are:

The Politics of Excellence and Choice in Education
William Boyd and Charles Kerchner (1987)

The Politics of Reforming School Administration
Jane Hannaway and Robert Crowson (1988)

Education Politics for the New Century
Douglas Mitchell and Margaret Goertz (1989)

Politics of Curriculum and Testing
Susan Fuhrman and Betty Malen (1990)

The Politics of Urban Education in the United States
James G. Cibulka, Rodney J. Reed, and Kenneth K. Wong (1991)

The New Politics of Race and Gender
Catherine Marshall (1992)

The Politics of Linking Schools and Social Services
Louise Adler and Sid Gardner (1993)

A List of Editors and Contributors

Jay D. Scribner holds the Hartfelder/Southland Corporation Regents Chair at The University of Texas at Austin. His teaching interests focus on educational politics and policy at federal, state, local and organizational level of public school and higher education systems. He was editor of the NSSE Yearbook, *The Politics of Education* (1977), co-editor (with D. Layton) of *Teaching Educational Politics and Policy* (1989), and co-editor (with C. Marshall) of *The Micropolitical Behavior of Principals* (1991). He is a founding member of PEA and is one of its past presidents.

Donald H. Layton is Associate Professor Emeritus at the University of Albany, State University of New York. His continuing research specializations are the state politics of education, intergovernmental relationships in education, and religion, politics, and education. He is the current PEA historian, a former editor of the *Politics of Education Bulletin*, and a former PEA president. Professor Layton and PEA President Robert Wimpelberg were the recipients of PEA's first Distinguished Service Awards in April 1994.

Gary L. Anderson is Associate Professor of Educational Administration at the University of New Mexico. He is the author of numerous articles on critical theory, critical ethnography, and Latin American education. His recent books include (with K. Herr and A. Nihlen) *Studying Your Own School: An Educator's Guide to Qualitative Practitioner Research* (1994) and (with J. Blase) *The Micropolitics of Educational Leadership: From Social Control to Social Empowerment* (1995).

William Lowe Boyd is Distinguished Professor in the College of Education at Pennsylvania State University. He is the author or editor of numerous books and articles on educational policies and administration, including several comparative studies and analyses, and he is co-editor of the forthcoming 1995 PEA Yearbook on the New Institutionalism. In 1994, Professor Boyd received one of PEA's first two Stephen K. Bailey awards for his scholarly contributions to the politics of education.

James G. Cibulka is Professor of Administrative Leadership and Director of the PhD Program in Urban Education at the University of Wisconsin–Milwaukee. He is also Editor of *Educational Administration Quarterly*. His specializations include educational politics, policy, and finance, with particular attention to urban issues, private schools, and intergovernmental relations. His most recent book (with R. Reed and K. Wong) is *The Politics of Urban Education in the United States* (Falmer, 1992).

Robert L. Crowson is Professor of Education in the Department of Educational Leadership, Peabody College, Vanderbilt University. His major research interests include large-city school administration, the school principalship, the school district super-intendency, and the coordination of children's services. He has authored, co-authored or co-edited volumes on the principalship, organization theory, the politics of administrative reform, and school–community relations.

Frances C. Fowler is Assistant Professor in the Department of Educational Leadership at Miami University (Ohio). Her specialties in research and teaching are the politics of education and educational policy. She has published articles in both areas and also in comparative educational policy. She is currently secretary of the Politics of Education Association and a member of the Editorial Board of *Educational Administrative Quarterly*.

Lance D. Fusarelli is a doctoral student in the Department of Educational Administration at The University of Texas at Austin. His research interests focus on the history and philosophy of school reform, the politics of democratic participation, and educational politics and policy.

Laurence Iannaccone is Professor Emeritus at the University of California, Santa Barbara. He is the author of numerous books and articles on the politics of education, including *Politics in Education* (1967), (with P. Cistone) *The Politics of Education* (1974), and *Educational Policy Systems* (1975). Professor Iannaccone and Frederick M. Wirt were recognized with PEA's first two Lifetime Achievement Awards at the Association's 25th anniversary convocation in April 1994.

Frank W. Lutz is Professor of Educational Administration at East Texas State University where he directed the Center for Policy Studies for seven years. He has written extensively in the politics of education often with L. Iannaccone (1968, 1978). His most recent book, with C. Merz, is *The Politics of School Community Relations* (Teachers College Press, 1992). He taught earlier at New York University, Pennsylvania State University, and Eastern Illinois University.

Betty Malen is Professor in the Department of Education Policy, Planning, and Administration at the University of Maryland in College Park. Her specialization is educational policy, politics and administration. Her research has addressed the politics of prominent reform initiatives, notably site-based management, career ladders and tuition tax-credits deductions. In addition to publishing on those areas, she co-edited (with S. Fuhrman) *The Politics of Curriculum and Testing* (Falmer Press, 1991).

Catherine Marshall is Professor of Educational Administration at The University of North Carolina at Chapel Hill. She is the editor of the 1992 PEA Yearbook, *The New Politics of Race and Gender* (Falmer Press). She has published extensively about politics of education (from state to micro-level), qualitative methodology, women's access to careers, and about the socialization, language, and values in educational leadership. She is author of *Culture and Education Policy in the American States* (Falmer Press), and is a past PEA president.

Tim L. Mazzoni is Professor of Educational Policy and Administration in the College of Education at the University of Minnesota. His major teaching and research interests focus on educational politics and policy. Among his scholarly publications are 'Analyzing state school policymaking, an arena model' (Summer 1991) and 'The changing politics of state education policymaking: a 20-year Minnesota perspective' (Winter 1993), both appearing in *Educational Evaluation and Policy Analysis*.

Paul E. Peterson is Henry Lee Shattuck Professor of Government at Harvard University. He formerly taught at The University of Chicago and Johns Hopkins and was Director of Governmental Studies at the Brookings Institution. His book, *School Politics, Chicago Style* (1976), received the Gladys Kammerer Award of the American Political Science Association. For scholarly contributions to the politics of education Dr Peterson was a 1994 recipient of PEA's Stephen K. Bailey Memorial Award.

Pedro Reyes is Associate Professor of Educational Administration at The University of Texas at Austin. His research interests include the analysis of teacher work conditions and school-level policy making; he also studies the effects of state policy making on student outcomes. In 1992–93, he held a fellowship awarded by the National Academy of Education.

Norman Robinson is an Emeritus Professor of educational administration at Simon Fraser University in Vancouver, Canada. His research work has focused on political behavior in education and issues in educational governance. He has served extensively as a consultant on administration and governance matters in education to a number of governmental and nongovernmental institutions in Canada and abroad.

Kent Paredes Scribner is Assistant-Principal at Sunland Elementary School, Roosevelt District, in Phoenix, and a recent Graduate Associate in the Division of Educational Leadership and Policy Studies, Arizona State University. His research interests include multicultural education and the sociopolitical dynamics of parents and shared controls in schools.

Gerald E. Sroufe is Director of the Governmental and Professional Liaison Program at the American Educational Research Association in Washington, DC. His research interests center on state and federal politics of education. His most recent chapter on the politics of federal research policy appears in *National Issues in Education* edited by J. Jennings.

Robert T. Stout is Professor, Division of Educational Leadership and Policy Studies, Arizona State University. His research interests have been around questions of education and politics with particular interest in questions of equity.

Marilyn Tallerico is Associate Professor in the School of Education at Syracuse University. Her professional interests center on school board research and gender issues in educational leadership. She recently co-edited the book, *City Schools: Leading the Way* (Corwin Press). She was principal author of the monograph, *Gender and Politics at Work: Why Women Exit the Superintendency* (National Policy Board for Educational Administration).

Richard G. Townsend is Professor of Administration and Coordinator, Political Action, at the Ontario Institute for Studies in Education in Toronto. After work in civil rights, he now specializes in Canadian studies, writing about the beliefs of politicians, the rhetoric of administrators, the interest groups of teachers, and the agendas of students. He is the author of *They Politick for Schools* (OISE), and is the founding editor of *Tales from the School*.

Tyll van Geel is Taylor Professor of Education at the Warner Graduate School of Education and Human Development, University of Rochester. His specialization is educational law and policy, but he also teaches courses in educational politics and school administration. He has written numerous books and articles, including *Educational Law* (McGraw-Hill) with M. Imber.

Kenneth K. Wong is Associate Professor in the Department of Education and the College at The University of Chicago. He is author of *City Choices: Education and Housing* (SUNY Press) and co-editor of *Rethinking Policy for At Risk Students* (McCutchan) and of *The Politics of Urban Education in the United States* (Falmer Press). He is Principal Investigator of a research project, 'Systematic Governance in the Chicago Public Schools.'

Foreword to The Study of Educational Politics

Paul E. Peterson

The politics of education is both a critical and a conservative discipline. It is skeptical of the institutions that it studies, but it is cautious about proposing changes. If its treatment of those in authority is often sardonic, it is no less dubious of proposed remedies and reforms.

Most of the scholars working in schools of education devote themselves to the training of practitioners – the teachers, administrators, counselors, psychologists, and special educators who staff this multi-hundred billion dollar industry. Almost necessarily, they come to identify with the needs, values and interests of the enterprise with which they are closely identified. They typically defend the professional against the laity, the expert against the novice, the specialist against the generalist, the insider against the external critic. Yet from time to time educational scholars reverse fields and propose and help launch wholesale reforms of one or another part of the educational enterprise.

Not so are most scholars who write about the politics of education. Their skeptical conservatism is deeply rooted. The first modern political analyst, Niccolo Machiavelli, was the quintessence of ruthless cynicism. He advised the Italian princes of the late 16th century not to pursue justice but to keep themselves in power. In pursuit of this goal, they should impose necessary harms immediately upon coming to power, then let the benefits of the prince's rule dribble out gradually over time. The public would soon forget the harms and become grateful for their subsequent benefits. He also recommended memorable executions of opposition leadership both to deter guerrilla action and to win public awe and respect.

The first great modern political theorist, Thomas Hobbes, who wrote in the aftermath of England's great Civil War, was willing to concede all authority to a single all-powerful sovereign in order to avoid a state of nature that was 'nasty, brutish and short.' Anything other than absolute despotism, he said, would inevitably degenerate into mob rule. Government was not so much a positive good as a necessary evil.

James Madison, the founder of the American political tradition, was hardly less restrained in his enthusiasms. He advocated the separation of powers among competing branches within a federal system not so much so as to achieve good government as to keep any one faction from gaining power. The balance of power among competing interests was the only way to maintain liberty.

The writers of the essays that follow are steeped in this tradition. They neither pander to those in authority, nor do they endorse recommendations made by critics of right or left. They neither portray American education as controlled by mindless multi-culturalists and self-serving bureaucrats nor criticize it for anti-democratic élitism. Instead, schools are seen as a strategic battleground over which many a brigade has marched and on which are buried the bodies of many a combatant.

It is no accident that the politics of education established itself as a field in the late 1960s. At that time, many thought social problems could be solved through political action. School boards were thought to be vehicles by which a white majority suppressed

0268-0939/94 $10·00 © 1994 Taylor & Francis Ltd.

minorities. If the boards could be changed, so could the schools. School administrators were condemned for their insulation from a multiplicity of political interests. If bureaucrats could be subjected to political influence, better schools could be created. Claims to professional expertise were subjected to harsh scrutiny. If schools could be made responsive to the laity, children would learn more. Some even embraced participation of the poor in the hope of bridging the distance between the school values and those taught in the home.

But if these were the circumstances that induced schools of education to accept the politics of education as a new discipline, those contributing to this emerging discipline adopted a skeptical stance toward much of what was happening around them. To be sure, they agreed with those social critics who accused schools of being controlled by small factions. Yet only a very few were ready to join hands with the civil rights groups, citizen advocates, policy analysts and educational reformers who called for the restructuring of American education. Too steeped in a tradition founded by Machiavelli, Hobbes and Madison, any espousal of reform was tempered by a keen awareness that more than one group or faction wore a self-interested hat.

Looking back on the past 25 years, one can appreciate the virtue of cautious skepticism. Educational governance has changed markedly, especially in large American cities. People of color have gained representation; teacher organizations have become an integral part of the governing structure; states have become more involved in school finance; and federal regulations limit the authority of local school boards. But for all these changes in the politics of education, the productivity of the educational system itself seems to have changed but little. If American schools have not deteriorated as much as reform commissions have claimed, neither have they found new ways of enhancing learning opportunities. If the system seems more stagnant than rotten, the result can hardly be comforting to yesteryear's enthusiastic reformers.

And so the stance of those contributing to the politics of education remains cautiously critical. Educational reform remains as much a part of the politics of the 1990s as it was of the 1960s. Some reformers ask that parents be given vouchers so that they may have their choice of school, public or private, religious or sectarian. Other reformers ask that control of schools be decentralized to neighborhood groups. Still other reformers think that equalizing school expenditures will level the playing field. Testing teachers, chartering schools, contracting school systems out to private firms, and 'detracking' classrooms: these and many other innovations remain on the political agenda.

These reforms are not rejected out of hand in the essays that follow. Skeptical conservatism does not preclude a willingness to explore, experiment, and innovate. But cautious skeptics moderate claims that educational reform can bring panaceas, permanent solutions, or a politics without problems. The reader is to be forewarned: what follows is political wisdom, not educational whimsy.

Introduction and overview

Jay D. Scribner and Donald H. Layton

In early 1993, the publications committee of the Politics of Education Association stipulated that the focus and content of the 1994 PEA Yearbook should commemorate the Association's 25th anniversary in 1994. The Yearbook, the committee stated, should chronicle the growth of politics of education scholarship, should document some of the forces that have shaped its development, and should assess the prospects for political studies of education in the years ahead. This volume, *The Study of Educational Politics*, is the result of the committee's call.

The Yearbook project has proved to be a challenging one both to its editors as well as to the contributors. In the 1960s, when the politics of education was emerging as a distinctive field, the research literature was still sparse. State-of-the-art papers and literature reviews could aspire to be somewhat comprehensive and could state 'what is known' with some reasonable assurance of accuracy. Obviously the emergent field of the 1960s is in the mid-1990s a rich and maturing field of research and study, albeit disparate, fragmented, and perhaps at times even schizophrenic. The editors have found that how one captures and depicts this mixed bag of thought and research between the covers of one book has not been an easy task.

Twenty-five years ago, the politics of education had its principal moorings in political science. The approaches and paradigms of that discipline (like behavioralism and pluralism) had a powerful impact on shaping early studies and research questions in the politics of education. Most political scientists who were early students of the politics of education (such as Bailey, Minar, and Zeigler) were schooled in these traditional political science approaches. This Yearbook's content demonstrates just how far that the politics of education has moved from its origins but, to be fair, the discipline of political science has itself changed dramatically during the same time frame.

There is little doubt that the study of educational politics has been greatly enriched by the addition of new and multiple perspectives and by the new critical approaches in the examination of its phenomena. But as some authors suggest in the following pages of this yearbook, there may have been some losses in the neglect of traditional foci and approaches. As Sroufe points out in Chapter 5, putting more 'politics' back into studies of the *politics* of education may be a needed and even necessary corrective at this stage of the field's evolution. Certainly we must continue the quest for the discovery and the understanding of those elements which give definition, purpose, and clarity to studies of the politics of education.

Overview of the Yearbook

The Study of Educational Politics consists of four parts or sections. Part 1 of the yearbook is entitled 'Values and origins.' In Chapter 1, Robert Stout, Marilyn Tallerico, and Kent

0268-0939/94 $10·00 © 1994 Taylor & Francis Ltd.

Scribner discuss 'Values: the "What?" of the politics of education.' They assert that values and ideology are central to the politics of education, yet are often neglected in research and writings. Kenneth Wong follows with Chapter 2, 'The politics of education: from political science to interdisciplinary inquiry.' Wong's essay anticipates much of the later content of the yearbook, especially the chapters of part 3. Part 2, 'The political arenas,' examines some traditional areas or foci of politics of education research. Laurence Iannaccone and Frank Lutz lead off with Chapter 3, 'The crucible of democracy: the local arena.' Tim Mazzoni elucidates 'State policymaking and school reform: influences and influentials' in Chapter 4. Next, in Chapter 5, Gerald Sroufe, a full-time practitioner of educational politics, addresses the topic, 'Politics of education at the federal level.' In Chapter 6, Francis Fowler examines perhaps the most problematic (from a definitional and research point of view) arena in 'The international arena: the global village.'

Part 3, entitled 'Emerging research directions and strategies,' explores modes of inquiry and approaches which have had or are likely to have considerable consequences for politics of education research. James Cibulka, in Chapter 7, dissects 'Policy analysis and the study of the politics of education.' Then, William Boyd, Robert Crowson, and Tyll van Geel explore 'Rationale choice theory and the politics of education: promise and limitations' in Chapter 8. A rapidly emerging area of study is elaborated on in Betty Malen's contribution, 'The micropolitics of education: mapping the multiple dimensions of power relations in school polities' (Chapter 9). Catherine Marshall and Gary Anderson then conclude part 3 with a consideration of 'Rethinking the public and private spheres: feminist and cultural studies perspectives on the politics of education' (Chapter 10). Two chapters constitute the yearbook's fourth and final part entitled 'A quarter-century in perspective: implications for future research.' In Chapter 11, 'Making the politics of education even more interesting,' Richard Townsend and Norman Robinson challenge a number of conventional insights and findings in the politics of education. Finally in Chapter 12, Jay Scribner, Pedro Reyes, and Lance Fusarelli draw upon some previous discussions and metaphors in the yearbook to address: 'Educational politics: and the game goes on.'

The editors hope that the 1994 Yearbook will make a useful addition to the politics of education research and reference literature. We especially desire that the volume be of direct value to current and future students of the politics of education and to their instructors. We also believe that the yearbook will serve as a fitting introduction to the study of educational politics for those whose principal discipline and research interests may lie elsewhere, yet seek some understandings of the politics area. And we trust that we are not immodest in suggesting that we hope that scholars of the next century will find this volume an insightful account and retrospective of the intellectual journey of a small but vibrant scholarly field.

PART 1
Values and origins

1.   *Values: the 'What?' of the politics of education*

Robert T. Stout,
Marilyn Tallerico, and
Kent Paredes Scribner

The purpose of this chapter is to examine how divergent values and belief systems are brought to bear in the politics of education. It will also illustrate the areas of schooling which seem most vulnerable to conflicts produced by opposing value systems, examining those areas over extended periods of history in the United States. We believe that in doing so we can highlight the content of the political struggles which scholars in the politics of education work to understand: the 'What?' of the field.

Appropriately, we begin with Easton's (1965) definition of a political system as:

> ... patterns of interaction through which values are allocated for a society and these allocations are accepted as authoritative by most persons in the society most of the time. It is through the presence of activities that fulfill these two basic functions that a society can commit the resources and energies of its members in the settlement of differences that cannot be autonomously resolved. (p. 57)

We could have begun as easily with Iannaccone's (1988) discussion of the place of political ideology in political conflict. He argues that:

> ... conflicts which escalate into realignment of coalitions and a redirection of policies, are reflections of 'intrinsically unresolvable issues about ... fundamental tensions inherent in American society.' Because continued political conflicts about such issues are likely to destroy a society, a substitution of conflicts takes place around a different mix of issues which promises a future solution to the problems posed by irreconcilable tensions. But precisely because they are irreconcilable, at least within the limits of their current circumstances and technology, the new mix of issues and related ideas provides an illusion of solving the old conflicts. (p. 58)

We will examine five questions which seem to us to have been the 'What?' of the politics of education throughout the history of public schools in the United States:

- Who should go to school?
- What should be the purposes of schooling?
- What should children be taught?
- Who should decide issues of school direction and policy?
- Who should pay for schools?

These questions continue unresolved, only having been decided one way or another at one time or another. We argue that the questions are unresolvable because they rest on underlying tensions among competing values. In other words, they cannot be resolved in a pluralist democratic system. The value tensions we will explore are linked to the value alternatives of choice, efficiency, equity, and quality (excellence), and we will argue that the tensions have surrounded public schooling since its invention in the United States. What people have fought about and what scholars of the politics of education have tried to understand are the ways in which major actors with competing value perspectives have tried to impose their perspectives on social policy. While a great deal of research in the past 25 years has been devoted to tracking the outcomes of conflicts over the five

0268–0939/94 $10·00 © 1994 Taylor & Francis Ltd.

questions, we hope to broaden the context with a longer historical perspective.

We intend, as well, to review these questions and their underlying value tensions within the research arenas of micropolitics, school district politics, state politics, and national politics. We do so with the understanding that other chapters in this yearbook will examine each arena in much greater detail.

It is worth noting, as a preface, that the inclusion of an entire chapter on this topic in a volume of this sort is a relatively new phenomenon. It is not that values have not been lurking in the background of the politics of education, but rather that a direct examination of their influence on political processes and outcomes is recent. Much of the politics of education research intended to illuminate the structures, actors, and processes of political decision. The value content of the issues was less well analyzed.[1]

One of the more widely used textbooks, *Schools in Conflict* (Wirt and Kirst 1992) does include a chapter devoted to the origins of demand inputs. The authors argue that stress in political systems arises from value conflicts among competing political agents. They then discuss the four key values of quality, efficiency, equity, and choice. They quote extensively from works by Marshall, Mitchell, and Wirt, citing particularly *Culture and Education Policy in the American States* (1989). In that work, after giving credit to Kaufman (1956), and Garms, Guthrie and Pierce (1978), Marshall *et al.* elaborate their terms. On the definition of *choice*, they say:

> This is arguably the most basic of all American public values. It was the passionate belief of the American Federalists that good government is defined by its ability to preserve freedom of choice for its citizens . . . It was summed up by Thomas Jefferson in his declaration, 'That government governs best which governs least'. (p. 89)

About *quality* they say:

> Given the primary role played by choice or liberty in the American political system, positive public policy actions must be justified in terms of their ability to enhance the quality of life for citizens. Indeed, governmental action to provide direct services is defensible only if the quality of the services provided is on the whole at least as good as could be reasonably expected to arise through private action. (p. 90)

With regard to *efficiency* they explain that:

> Americans have had an intense love–hate relationship with efficiency as a public policy value since the founding of the Republic. The cruel efficiencies of totalitarian governments are recognized and feared. But the productive efficiency of American business and industry are just as frequently held out as a model after which to design public service agencies. Moreover, Americans feel a need for an orderly, predictable, and controlled system to contain private and interest group conflicts threatening the social order. Social unrest and the threat of anarchy fade when government provides for the orderly and efficient delivery of public services. (p. 91)

About *equity* they argue:

> As a policy matter, equity is complicated. It is a matter of *redress* rather than one of *address*. That is, policy-makers cannot decree social equity, they can only create laws and social programs that relieve the effects of inequity *after* it has been identified. The need for governmental action cannot be recognized until some identifiable inequity has been shown to be serious and in need of remedy. Then action is only justified to the extent necessary to eliminate the identified inequity. (p. 92)

We return, then, to Iannaccone (1988) who asserts, 'Policy is thus viewed as resting upon value-laden public beliefs – interpretations of the American creed or dream – as you will' (p. 49) for our interest in examining the connections among values and political processes and outcomes.

Who should go to school?

Although we risk overgeneralization, we assert initially that this question has been driven over time by a shift from 'choice' to 'equity' as the dominant public value influencing decisions about it. While we can find undertones of both 'efficiency' and 'quality/excellence,' the overall drift of public policy in the past 200 years has been to reduce individual choice in favor of social equity. At the base of the question are individual choices to go or not to school, to choose the persons with whom one wishes to go to school, and to choose the persons with whom one does not wish to go to school.

At the beginning, of course, the New England colonies defined the terms of resolution in favor of excellence, arguing that public provision of schools was essential for the quality of life of all persons in the social order. But implementation was a sometime matter, resting essentially on town choices to allocate funds. In other regions of the nascent country, choice was the dominant early value. The framers of the Constitution avoided what they saw as a series of potential conflicts by avoiding discussion of the question altogether. This is not to argue that children did not go to school, only that parents made the choices (perhaps under duress from religious leaders, or 'liberals', or business leaders). With the advent of compulsory schooling, choice gave way to equity as the dominant public value. But the conflict has lasted a long time, with compulsory school attendance laws still being debated and with some states only recently deliberating appropriate school-leaving ages.

With the rise in the salience of equity, the decision to expand (demand) school-going to everyone seems inevitable in hindsight. But conflict over the meaning of the question has been intense: witness conflict over school segregation, education of handicapped children, education of Native Americans, and the like. Although we believe equity has been the dominant value for at least 100 years, choice continues to influence policy. Whether parents can provide schooling at home, whether children who are violent or truant must be schooled, whether homeless children are entitled to schooling, are all issues which can stir substantial debate even now. With respect to higher education, the shift from federal grants and fellowships to student loans indicates that choice may have reappeared as a more powerful value than in the recent past.

In the micropolitical arena, the question of who should go to school has not had substantial investigation by scholars of the politics of education. Rather, sociologists, particularly those who have been concerned about such matters as class and race and the interactions between students and teachers, have had more to say about it. But there are examples of research which suggest that a more direct look at political ideology might help explicate internal school politics. In studying high school dropouts, Reyes and Capper (1991) explore the political ideologies of a sample of urban principals. Principals, they assert, determine in part the nature of dropping out by how the principals define *dropout* and what proximate causes they assign. Reyes and Capper argue, 'In sum the principals blamed the student, the school or community context, for the dropout of racially diverse students. None attributed student dropout to reproducing the status quo within society . . .' (p. 549). They conclude by saying, 'In summary, our findings confirm . . . that how a problem is defined can determine if and how the problem is addressed' (p. 551). In effect, how the problem is defined at a school determines who is to go to school.

More recent compendia have made similar arguments with respect to problem definition (issue articulation) research which might be carried out at the micropolitical level. Blase (1991) argues that, while micropolitical processes are complex and unstable, school principals have much to do with problem defining. He argues that in an effort to achieve a deep awareness of self, school principals must examine their own political values and purposes, and assess the political values of others.

A recent issue of *Education and Urban Society* edited by Marshall and Scribner (1991a) is devoted to micropolitics. It is clear from this volume that conflict and accommodation at school sites will be a topic of research for many future scholars. We suggest that some of the investigative effort be devoted to uncovering value conflicts among the participants over the key issue of who is to be a student and who is not to be. We assert that conflict over the core values of choice, equity, efficiency, and quality exists routinely in and around schools in their attempts to decide who should go to school, notwithstanding the general dominance in national affairs of the value of equity.

The question of who should go to school has been hotly contested in school districts in the past 25 years. Major conflicts have centered on issues of race and school desegregation (Crain, 1968, 1989, Kirby *et al.* 1973, Willie 1984). Political scientists have examined the political structures and processes that drove decisions which did or did not desegregate schools and have examined the consequences in terms of white flight, resegregation, and housing patterns. As Crain (1989) argues, 'In our analysis of education, we rarely consider that the local school system is a very powerful actor – it is a major employer and builds most of the city's buildings' (p. 318). Other emerging district issues include treatment of truant and whether city police will apprehend them, integration of social service delivery systems (Melaville and Blank 1993), questions of academic qualifications as requisites for participation in school-sponsored activities (pass to play), and youth violence.

Here we believe that the core value of equity is in contest with values of quality and efficiency. The costs in time, money and lost academic performance are weighed against the expressed obligation to give every child in a system an equal opportunity to succeed. It is not clear to us how these contests will develop nor is it clear that current work in the politics of education is sufficiently advanced to explain them, although Bidwell (1992) has laid out the components of a possible scheme for thinking about urban education as a field of policy action. Further, Schwager and others (1992) have given us an analysis of the complex implementation effects of district policies about retaining children in grade. They argue that an interaction of district cultural beliefs and organizational procedures produces different rates of retention in grade depending on the size of the school district, even when the formal policy is the same. This produces different answers to the question of who should go to school, since grade retention is shown to be linked to dropping out.

State action on this question, lately, has been anchored in both efficiency and excellence (Firestone, 1990). Raising high school graduation standards, either by extending required courses, or by instituting some form of 'leaving' exam, is justified in terms of excellence, as is the 'pass to play' rule in some states. Providing state money for increased efforts at early intervention, either through the schools, or through other agencies in cooperation with schools, is justified in terms of long-term efficiencies.

Federal action has been in a slow drift toward equity, intensified, perhaps, in the past 25 years. For example, inclusion and equal educational opportunity are symbolic of a national mood which values schooling for all children and youth. Yet, the countervailing value of choice continues in the federal shift from grants to loans for college students, and in, for example, the US Supreme Court's reluctance to enter suits over school finance.

Overall, we suggest that the prevailing value behind decisions about who should go to school has been equity. Although periodic incidents and decisions have been flavored by efficiency and choice, most decisions most of the time in most political systems have reflected a preference for equity.

What should be the purposes of schooling?

This is, of course, the most significant question in the group, because, without a clear answer to it, the other questions are much more contentious. But, as a nation, we have not answered this key question. Having said that the purpose of public schooling is to advance the interests of the public as represented by the state, and to prepare a coming generation for success in the future, we have engaged in serious debate about what those ideas mean. The debate is not new of course, having its origins in the earliest proclamations. In 1642 the Massachusetts Bay Colony set down a simple purpose:

> It being one chief object of that old deluder, Satan to keep men from the knowledge of the Scriptures, as in former times by keeping them in an unknown tongue, so in these latter times by persuading from the use of tongues, that so at least the true sense and meaning of the orginal might be clouded by false glosses of saint-seeming deceivers, that learning may not be buried in the grave of our fathers in the Church and Commonwealth, the Lord assisting our endeavors . . .

Matters did not remain simple. The 1754 catalog of Queen's College (Columbia University) asserted that the 'chief thing that is aimed at in this College is to teach and engage the children to know God in Jesus Christ . . .' By 1784 the Constitution of the Commonwealth of Massachusetts asserted, 'Wisdom and knowledge, as well as virtue, diffused generally among the body of the people, being necessary for the preservation of their rights and liberties; . . .' Above the Boston Public Library is engraved, 'The Commonwealth requires the education of the people as the safeguard of order and liberty.'

In 1749, in Proposals Relating to the Education of Youth in Pennsylvania, Benjamin Franklin added to the debate:

> As to their studies, it would be well if they could be taught everything that is useful and everything that is ornamental. But Art is long and their time is short. It is therefore proposed that they learn those things that are likely to be the most useful and most ornamental. Regard being had to the several professions for which they are intended.

In 1848 Horace Mann was prompted to write,

> Now surely nothing but universal education can counterwork the tendency to the domination of capital and the servility of labor . . . It [education] does better than to disarm the poor of their hostility toward the rich; it prevents being poor.

What is/are the purpose(s) of education in the United States? President Bush (1991), in presenting America 2000: An Education Strategy, said:

> Education has always meant opportunity. Today, education determines not just which students will succeed, but also which nations will thrive in a world united in pursuit of freedom in enterprise. (p. 1)

> If we want America to remain a leader, a force for good in the world, we must lead the way in educational innovation. And if we want to combat crime and drug abuse, if we want to create hope and opportunity in the bleak corners of this country where there is now nothing but defeat and despair, we must dispel the darkness with the enlightenment that a sound and well-rounded education provides. (p. 2)

> Think about every problem, every challenge we face. The solution to each starts with education. (p. 2)

In the same document, the report of the meeting of the Governors of the several states and the President in Virginia at the education summit in 1990 is excerpted as:

> America's educational performance must be second to none in the 21st century. Education is central to our quality of life. It is at the heart of our economic strength and security, our creativity in the arts and letters, our invention

in the sciences, and the perpetuation of our cultural values. Education is the key to America's international
competitiveness. (p. 59)

What are the purposes of education? Over time they seem to be whatever we decide we
want them to be at the moment, aside from a general belief that national happiness is at
stake if they are not met.

We are not surprised, then, at the corollary debate over whether the schools are any
good (Berliner 1993, Hawley 1985, Timar 1989). As a nation we cannot possibly agree on
that issue, given our inability to agree on the prior question of purpose.

We wish we could sort out the valences of the competing values of choice, efficiency,
equity, and excellence in the debates over purpose. However we do not think that we can
within the scope of this chapter. We do not have the wisdom of hindsight except to assert
that the debates have been contentious.

What should children be taught?

In general, we believe that the question of what children should be taught has been
decided around the competing values of excellence and equity, although, on occasion, the
value of choice appears to have driven the debate. As is true for all of the important
questions, this one has been debated in various forms for more than 100 years. In the late
1800s the American curriculum can be described best as in disarray. Choice prevailed at all
levels in the absence of any general agreement about what should be taught. Teachers
taught what they knew, proprietary schools taught what they could sell, parents and
students demanded different curricula depending on social status, regionalism, religion and
the like. The Committee of Ten of the National Education Association (NEA 1894)
attempted to solidify the national curriculum around subjects which were thought to
prepare high school graduates for success in college. About 25 years later the Commission
on the Reorganization of Secondary Schools (1918) published *The Cardinal Principles of
Education*. This document argued, in effect, that the high school curriculum had to be
modified substantially to allow successful completion by large numbers of students who
had not been in high schools in the late 1800s. The debate was thus joined between those
who argued for excellence and those who argued for equity. Both groups had abandoned
choice as a preferred value. The debate is engaged in the same terms today.

At the level of micropolitics an uneasy compromise has been reached with the tacit
acceptance of various forms of tracking, so that excellence can be celebrated for the
children who are thought to be able to benefit, and some form of presumed equity offered
for less able students (Oakes 1985, Powell *et al.* 1985). While these practices may be racist
or discriminatory, they allow schools to function without continual rancorous conflict.
Other compromises are known to occur (O'Reilly 1988), but are not well documented as
outcomes of value conflicts. Obvious compromises include a school faculty agreement that
some teachers can teach phonics while other teachers are permitted to teach whole-word
approaches, or to teach reading through whole-language experiences, or to avoid teaching
some subjects altogether. The closed classroom doors and the loose internal coupling of
most schools permit value conflicts from surfacing. But value differences persist.

Within school districts, Boyd (1976, 1978) has reminded us how complex and
interesting the political contests have been over what should be taught. He has given us a
model for understanding how the contests are waged. Both Peshkin (1978) and Page and
Clelland (1978) provide vivid studies of the ways in which communities can mold what is
taught so that it reflects the dominant values of the community. Although we do not

know with certainty, it is reasonable to speculate that most local curricula reflect the values of key actors (Boyd 1982, 1987, Burlingame 1988), whether those values are equity, excellence, or choice. Probably few 'pure' cases exist, but Boyd's (1978) model might be useful for additional research on the problem.

There is, we believe, a resurgence of earlier overt conflicts, particularly in state-level political arenas. Marshall. Mitchell and Wirt (1989) proffer the most thorough explication of how such conflicts result in very different choices among states. Among their seven major policy domains which engage state legislators, three are about the question of what should be taught: approaches to student testing and assessment, approaches to curriculum materials, and approaches to the definition of school program. They show that states differ across the four values and across the seven policy domains. But individual states, they argue, have general preferences among the values. They argue also that there are general value preferences shared by all the states. On that point they say, 'It was surprising to find so little priority given to approaches that would enhance the choice value' (p. 94). They say also, 'Note, for example, that educational *quality items were ranked first in all...domains...'* (emphasis in the original: p. 93) and *'Receding support for educational equity is clearly evident in the data'* (p. 94: emphasis in the original).

The overt conflicts are perhaps best represented by state debates over high-stakes testing. As states attempt to attach serious consequences to various forms of the tests, the content of them, the cut scores for passing, and the consequences of failure. As Ellwein, Glass and Smith (1988) show, each of these debates is subject to various forms of political compromise at each juncture.

The more recent emergence of state conflict over the inclusion of values in Outcomes Based Education represents other instances (see the 1994 conflicts in Kentucky and Pennsylvania for example). We are not certain how these debates will continue, but the conflicts seem anchored in definitions of excellence and equity. Some attention is being paid to efficiency as states discover the high costs of new, and presumably more equitable, forms of testing. But states, as Boyd (1987) and Astuto and Clark (1986) argue, will probably continue to base policy on excellence as they debate the question of what students should be taught.

Federal political debates have the same flavor. Whether the issue is crystallized in multiple attempts to articulate the national goals (see for example the *Reaching the Goals* series produced by the various goals work groups of the Office of Educational Research and Improvement, US Department of Education, [1993 and various dates]), or by the multiple attempts to establish national curricular standards, the federal drive for excellence seems well established unless it becomes enmeshed in technical wrangling (US General Accounting Office, 1993). Iannaccone (1985) was one of the first to suggest that a sea change had occurred in federal political values, and that excellence had replaced equity as the preferred symbol.

Whitt, Clark and Astuto (1986) agree and argue that public preference for the excellence symbols is high and likely to remain so. The 1993 Phi Delta Kappa/Gallup Poll of public attitudes seems to reaffirm continuing public support for high curricular standards and the teaching of traditional values such as honesty, democracy, tolerance, patriotism and the like (*Phi Delta Kappan*, October 1993). This is not to argue that excellence as a preferred value has replaced equity entirely, only that equity now has to be viewed as a mitigating concern, rather than as the primary one, in federal debates about what children should be taught.

What is happening, then, is what has happened throughout our recent history. The American political debate over what children should be taught plays out at several levels, and continues to revolve around the competing values of equity and excellence.

Who should decide issues of school direction and policy?

This is a most interesting question, about which research in the politics of education has had much to say. Over the past 100 years or so some trends are apparent. As a country we have debated the appropriate role of citizens in the governance of schools. Governance of local school systems changed from a diffuse and decentralized mode in tha late 1800s to a generally centralized and professionalized mode by the 1950s. Since the 1960s modes of governance in local school systems have become more diverse, serious questions have been raised about the political health and viability of locally elected school boards, and new forms of governing are being tried. We cannot even speculate on the outcome, except to remind readers that somewhere in the debate lie competing values of efficiency and equity. Efficiency may be represented by those who argue that local school boards, as now constituted, are not up to the task of governing a complex modern school system. Those who favor equity insist that the public's right to govern its schools in whatever ways it sees fit, and in however confusing or messy a way, must be protected.

We have seen also the increased capability of both state and federal agencies to intervene in local school systems. Their interventions have changed the character of decisions about schools. Finally, we have come to understand that schools are political arenas as well, and that while influenced by districts, states and the federal government, teachers, principals, and parents contend with one another over who will be in charge.

At the micropolitical level we have an increased understanding of the rules of political conflict (Bacharach and Lawler 1980, Ball 1987, Blase 1988, 1991, Marshall and Scribner 1991b, Malen and Ogawa 1988). But we do not have a particularly good idea of the content of these conflicts, and whether the conflicts are about matters of political ideology. We suspect that they are, and perhaps more often than on occasion (Iannaccone 1991, Spring 1988, Wolcott 1977). Even in its original use by Iannaccone (1975), micropolitics is concerned 'with the interaction and political ideologies of social systems of teachers, administrators and pupils within school buildings' (p. 43). So we suspect that struggles among teachers, parents and administrators are about important issues, however they may be disguised. Iannaccone (1991) suggests that teasing out those issues may be difficult, but worth the effort in the context of research on reform.

The politics of governing local school districts has been the focus of much attention by researchers. From early in the development of the politics of education, inquiry into the question of who decides has, more indirectly than not, surfaced questions of values. Berube and Gittell (1968) offer views of a struggle that was over parent influence, but about deeper ideological differences. Levin and his colleagues (1970) explored in depth the issue of community control of schools. Fein (1970) placed the issue squarely in the middle of ideological conflict. He wrote, 'But it is when the issue is political-ideological reform that the debate sharpens . . .' (p. 86). Clarifying his perspective, he argued that the liberal critics of public schools were in a quandary over how to deal with particularistic ideologies as presented by African-American parents. 'These several [universalistic] beliefs are directly at odds with the theses now propounded by defenders of community schools' (p. 90).

Mitchell (1974, 1980) has argued that school board members value differentially, and that their values are important to the outcomes of policy debates. Crain (1968) suggests that the values of school board members may have been one of the key variables explaining the outcome of community conflict over school desegregation. Others (Cahill 1964, Kimbrough 1964) have made the same point. In fact, Cahill (1964) wrote, 'In the second place the political values of the participants encouraged them to select particular *kinds of political change* [emphasis in the original] for attention. In this case, differences in the value perspectives of the participants generated corresponding differences in patterns of selection' (p. 68). Heineke and Brand (1994) have completed careful analyses of the public speech of school board members, and have discovered conflict which arises out of value differences among incumbent school board members.

We can say with some certainty, then, that conflict over who should decide in local communities is important because it is conflict over whose values will influence school policy. Certainly the work done by those who use dissatisfaction theory begins there. These researchers (Chriswell and Mitchell 1980, Danis 1981, Iannaccone and Lutz, 1970, Lutz and Iannaccone 1978, Weninger and Stout 1989) have attempted to understand how shifts in dominant community values produce electoral conflict over school board seats, replacement of school superintendents, and major changes in policy directions in school districts.

Although we cannot be certain, we can speculate that the core values discussed in this chapter influence local political decisions in at least two ways. It may be that community values (equity, efficiency, for example) play an important role in structuring the rules of political conflict. It may also be that the values influence actions of key players, independent of the rules of political conflict (Boyd 1976).

The work of Tallerico (1989) and others (Cistone 1982, Danzberger *et al.* 1986, Danzberger *et al.* 1992, Lutz 1984, Lutz and Gresson 1980, Stout 1982) all suggest that the rules of political conflicts in local school districts are established as a function of community values articulated in various ways. Whether the values we have been discussing are the best ones to assess in trying to understand the complex political life of school districts is unanswerable in this space. Tallerico (1989) suggests that the push and pull of activist school board members interacting with school superintendents 'are powerful predecisional social processes that create the conditions and shape the choices of alternatives upon which policies and practices are constructed...' (p. 227).

At the state level, values appear to influence the rules of the game in much the same way. State legislatures may be open or closed, public- or private-regarding, accessible or inaccessible, structurally complex or simple, reliant on staff or not, and show differences in a variety of means and methods for controlling the flow of policy debate (Fuhrman and Rosenthal 1981, Marshall *et al.* 1989, Mazzoni 1993, Mitchell 1988, Stout 1986). Whether a reflection of the political cultures of the states or other variables is not clear, but it is clear that state policy makers are influenced by values when establishing rules of operation. So issues of who should decide are first decided by values which govern the debate about who can even be part of the debate. The most recent development has been the enthusiastic reinclusion of business leaders in framing state education policy (Proseminar on Education Policy 1991, Ray and Mickelson 1990).

State legislatures have, as well, worked to define the question of who should decide by attempting to change the structures of local decision-making. As creatures of the states, local school boards are inventions of legislatures, even though many local school boards predate admission of states to the Union. A variety of mechanisms has been proposed to break what many see as the obstructionist stranglehold of local school boards. Charter schools and vouchers are only the most recent efforts by state legislators to bypass school boards and put decisions more directly in the hands of parents, teachers, or both. Other mechanisms may become attractive as well (Danzberger *et al.* 1992) if charter schools and vouchers become politically too difficult. Whether prompted by efforts to increase equity, excellence, or efficiency, choice seems to be the public symbol attaching to the proposals.

The federal government's attempt to define the question of who should decide needs to be analyzed by governmental branch. The courts seem to have been much interested in ensuring the rights of children and parents to participate significantly in decisions about schooling. Decisions which constrain professional discretion and expand student and parent discretion have come down with regularity in the past 25 years. The most obvious examples are drawn from decisions about the rights of children with handicapping conditions and who is to define an appropriate educational environment for them. We interpret these as decisions grounded in equity, and beginning certainly with the *Brown* decision.

Congress seems to be moving in several directions. Beginning with the passage of the Elementary and Secondary Education Act, parent participation (an equity value, we believe) in school decisions has been promoted. Recent efforts to promulgate national curricular standards, prompted by efforts to establish national goals for schooling, seem anchored in concerns for excellence. But the obvious trade-off is argued to be with respect to equity if the curriculum standards are ethnocentric, or biased in other ways. A second trade-off is argued to be with choice. If schools are to adopt national standards and a national curriculum, buttressed by national tests, the argument is that the basic choice of parents to decide what children will be taught will have been eroded. By helping to promote the New American Schools Development Corporation and other innovations in school design, Congress also seems to be promoting both choice and excellence. Thus, it is hard to determine whether any value is predominant in Congress as a body.

The Executive Branch, certainly since the election of President Reagan and mitigated only partially by President Clinton, seems solidly to favor excellence and choice. Clark and Astuto (1986) have described these developments in detail.

Overall, the federal government seems increasingly willing to suggest that it should have a significant role in decisions about important matters. By establishing standards and the frameworks for debate at other levels, the federal government has substantial influence over who decides.

Who should pay for schools?

As is true with other important questions, we assert that over 200 or so years of our history, the question of who should pay for schools has been generally driven by political values. In the case of finance, we believe that equity has been the preferred value. Although the earliest laws placed the burden for financing schools on the commonwealth, practice was far different. The schooling of children and youth has only gradually (in the long term) come to be accepted as a general public responsibility. Earlier efforts to fund schooling through parent obligations, lotteries, philanthropies, and the like gave way, over time, to levied taxes. Although Guthrie (1988) warns, 'It is virtually impossible to predict the valence of public concern for a policy-related value at a particular point in the future' (p. 386), we think that equity will continue to be a powerful force influencing the answer to the question. But aside from that very broad, and admittedly risky, prediction, debate over who should pay has intensified in the past 25 years.

At the micropolitical level, there is virtually no research to suggest the criteria used within a school to determine the non-budgeted source and use of funds. Bake sales, teacher purchases of materials, entrepreneurial principals, gifts from parents, student councils, friends of the school, and business partners are all means and sources of extramural funds. Research by scholars in the politics of education is needed to understand how allocations of these revenues are made, both within schools and, either formally or informally, between schools. Although weighed against budgeted funds, such revenues are probably not significant. A story in *Education Week* indicates that the Council for Aid to Education estimated the value of gifts and services in 1990–91 to public schools by foundations, corporations, and individuals to be $300 million, or less than 1% of annual school expenditures (*Education Week* 1992). Nonetheless they may represent important advantages to certain kinds of school communities, and the value of such gifts may rise in the future.

Mostly under threat of legal action, state debates about who should pay are driven by questions of equity, with some attention paid to questions of efficiency. Excellence is given symbolic prominence, but not sustained financial support. Whether adequacy (as a proxy for quality) can become grounds for either judicial or legislative action is undetermined, but adequacy has begun to appear as a political symbol (Jordan and Lyons 1992). The value preferences of legislators change, however, as allocation, rather than aggregation, decisions are made. Allocation decisions, it is argued, are driven first by quality, then by efficiency, and then by equity (Marshall *et al.* 1989).

The national value preference with respect to the question of who should pay seems to be the choice. Although the federal government's allocation decisions are not unlike those of the states, the question of who should pay has been decided for the most part by the courts. In the *Rodriguez* case (1973), the Court seems to have said that the federal interest in answering the question of who should pay for schools must defer to state decisions, thereby affirming state choice as the preferred value. The decision, of course, did not challenge the federal government's right to collect taxes and to disburse them to schools. Both efficiency and equity values underlie federal efforts to generate revenue. But the effects of court decisions are far greater than the small financial contribution made by Congress to public schools.

A question of some interest to future research in the politics of education may involve the politics of the delivery of integrated social services for children and youth. As argued in Gardner (1992) and Jehl and Kirst (1992), new forms of delivery of social services to youth are both possible and desirable, but problematic in their implementation. But implementation raises a set of questions about who should pay. These are both interagency and intergovernmental in character. Some services will be provided by municipalities or counties, paid for by general tax revenues; some may be provided by states. Some may be provided by school districts and some by the federal government. And so it can go, with various crosscuttings of sources of funds and mechanisms for delivering services. Thus the issue becomes much broader than that of who should pay for schooling. It becomes one of who should pay for general child welfare, particularly as child welfare can be shown to have significant influences over schooling. We believe that the politics of this question can be a focus of future research.

Although we cannot predict the value preferences of the future, public attitude may reflect a growing concern for equity. The 1993 Gallup Poll (Elam *et al.* 1993) suggests several threads of public concern for equity. In significant percentages, citizens favor allocating the same amount of money for all students, 'even if it means taking funding from some wealthy school districts and giving it to poor districts' (p. 142). They are moderately willing to pay more federal taxes to improve inner-city schools. They favor the provision by schools of a wide variety of social services. While citizens favor being able to choose schools within public school systems, they reject vouchers for private schools.

Stout (1993) contends that the issue is joined between those who argue for one or another free-market strategy and those who argue that strong government intervention is needed. Jennings (1992) has suggested that we may have emerged from about 20 years in which efforts to make schools better have not been accompanied by efforts to make them more equal. Although the general direction of the answer to this question has been to prefer equity, and although we have said that we believe it will continue to be so, we offer it as a large and inviting arena for research.

Concluding remarks

Lasswell has characterized politics as 'who gets what, when and how' (1936). Easton (1965) has depicted the political system as determining the authoritative allocation of resources and values for society. The politics of education ultimately resolves distributive questions in a material sense, as well as in terms of the citizenry's competing values, attitudes, and ideologies.

As Guthrie (1988) explains, 'The United States political system must accommodate individuals and groups whose values and belief systems at their roots often conflict with one another' (p. 373). In this chapter we have attempted to document the evolution of the politics of education by reviewing five questions which both: (a) reflect some of the most enduring value conflicts pertinent to education, and (b) capture the broadest concepts underlying scholarship in the field:

- Who should go to school?
- What should be the purposes of schooling?
- What should children be taught?
- Who should decide issues of school direction and policy?
- Who should pay for schools?

We have focused on the values of efficiency, quality, equity, and choice. Wherever possible, we have illustrated the conflicts surrounding each with reference to the work of politics of education scholars. And we have included examples from the political arenas of schools, school districts, statehouses, and the federal government.

In tracking the development of the field in this way, it becomes clear that the contested nature of these questions persists. Major issues are not settled, nor are major conclusions without controversy. The nature of 'good' education, who should govern, who should benefit, and how it should all be financed are questions whose answers are neither commonly understood nor agreed upon. Research on the politics of education has made substantial progress in unraveling the complexities of competing value systems and education, yet it is evident that our understandings of these interrelationships will remain incomplete. While we have taken a broad historical perspective on these issues, ensuing chapters revisit many of these questions with more specific attention to the past 25 years of politics of education scholarship.

Note

1. For example, Scribner and Englert (1977) did not discuss values of political ideologies in their introductory chapter of the NSSE Yearbook on politics and education. There is only one indexed reference to values in the Yearbook, and Iannaccone's (1977) chapter is the only one in which political ideology is discussed at any length. The more recent and massive bibliographic study of the field by Hastings (1980) does not index 'values' nor does it index 'political ideology'. Reading individual entries in the volume reveals that some attention was paid to the interaction of values and politics, but not as a direct question for examination. *The Handbook of Research on Educational Administration* (Boyan 1988) provides only one index reference to 'values' and that reference attaches to a discussion of models of organization, not to politics and education. The five chapters on politics and policy are essentially silent on political ideology as a force in educational politics. Our effort to place the contents of research in the politics of education within the framework we have chosen may, therefore, seem forced on occasion. We make no apologies for that.

References

ASTUTO, T. A. and CLARK, D. L. (1986) *The Effects of Federal Educational Policy Changes on Policy and Program Development in State and Local Education Agencies*, Occasional Paper Number 2 of the Policy Studies Center of the University Council for Educational Administration (Bloomington, IN: Indiana University).

BACHARACH, S. B. and LAWLER, E. L. (1980) *Power and Politics in Organizations: The Social Psychology of Conflict, Coalitions, and Bargaining* (San Francisco: Jossey Bass).

BALL, S. J. (1987) *The Micro-politics of the School: Towards a Theory of School Organization* (New York: Methuen).

BERLINER, D. (1993) Mythology and the American system of education, *Phi Delta Kappan*, 74(8): 632–641.

BERUBE, M. R. and GITTELL, M. (eds) (1968) *Confrontation at Ocean Hill-Brownsville* (New York: Praeger).

BIDWELL, C. (1992) Toward improved knowledge and policy on urban education, in J. Cibulka, R. Reed and K. Wong (eds) *The Politics of Urban Education in the United States: The 1991 Yearbook of the Politics of Education Association* (Washington, DC: Falmer Press) 193–199.

BLASE, J. (1988) The teachers' political orientation vis-a-vis the principal: the micropolitics of the school, in J. Hannaway and R. Crowson (eds) *The Politics of Reforming School Administration: The 1988 Yearbook of the Politics of Education Association* (Washington, DC: Falmer Press), 113–126.

BLASE, J. (ed.) (1991) *The Politics of Life in Schools: Power, Conflict, and Cooperation* (Newbury Park, CA: Sage).

BOYAN, N. (ed.) (1988) *The Handbook of Research on Educational Administration* (New York: Longman).

BOYD, W. (1976) The public, the professionals, and educational policy making: who governs?, *Teachers College Record*, 77: 539–577.

BOYD, W. (1978) The changing politics of curriculum policy-making for American schools, *Review of Educational Research*, 48: 577–628.

BOYD, W. (1982) Local influences on education, in H. Mitzel, J. Best and W. Rabinowitz (eds) *Encyclopedia of Educational Research*, 5th edn (New York: Macmillan and Free Press), 118–129.

BOYD, W. (1987) Public education's last hurrah? Schizophrenia, amnesia, and ignorance in school politics, *Educational Evaluation and Policy Analysis*, 9(2): 85–100.

BURLINGAME, M. (1988) The politics of education and educational policy: the local level, in N. Boyan (ed.) *Handbook of Research on Educational Administration* (New York: Longman), 439–452.

BUSH, G. (1991) Remarks by the President at the presentation of the national education strategy, *America 2000: An Education Strategy* (Washington, DC: US Department of Education), 1–3.

CAHILL, R. (1964) The three themes on the politics of education, in R. Cahill and S. Hencley (eds) *The Politics of Education in the Local Community* (Danville, IL: Interstate Printers and Publishers), 51–74.

CISTONE, P. (1982) School boards, in H. Mitzel, J. Best and W. Rabinowitz (eds) *Encyclopedia of Educational Research*, 5th edn (New York: Macmillan and Free Press), 119–130.

CLARK, D. and ASTUTO, T. (1986) The significance and permanence of changes in federal education policy, *Educational Researcher*, 15(8): 4–13.

COMMISSION ON THE REORGANIZATION OF SECONDARY SCHOOLS (1918) *The Cardinal Principles of Education* (Washington, DC: Department of the Interior).

CRAIN, R. (1968) *The Politics of School Desegregation: Comparative Case Studies of Community Structure and Policy Making* (Chicago: Aldine).

CRAIN, R. (1989) An update: The next step is theory, in J. Ballantine (ed.) *Schools and Society: A Unified Reader*, 2nd edn (Mountain View, CA: Mayfield), 314–318.

CHRISWELL, L. and MITCHELL, D. (1980) Episodic instability in school district elections, *Urban Education*, 15: 189–213.

DANIS, R. (1981) *Policy changes in local governance: the dissatisfaction theory of democracy*, unpublished doctoral dissertation, University of California, Santa Barbara.

DANZBERGER, J. KIRST, M. and USDAN, M. (1992) *Governing Public Schools: New Times, New Requirements* (Washington, DC: Institute for Educational Leadership.

DANZBERGER, J., KIRST, M., USDAN, M., CUNNINGHAM, L. and CARROL, L. (1986) *Improving Grass Roots Leadership* (Washington, DC: Institute for Educational Leadership).

EASTON, D. (1965) *A Framework for Political Analysis* (Englewood Cliffs, NJ: Prentice-Hall).

Education Week (1992) Gifts to educational institutions said to total $12 billion in 1990–91, *Education Week*, 5 August, 23.

ELAM, S. M., ROSE, L. C. and GALLUP, A. M. (1993) The 25th annual Phi Delta Kappan/Gallup Poll of the public's attitudes toward the public schools, *Phi Delta Kappan*, 75(2): 137–152.

ELLWEIN, M., GLASS, G. and SMITH, M. (1988) Standards of competence: propositions on the nature of testing reforms, *Educational Researcher*, 17(8): 4–9.

FEIN, L. (1970) Community schools and social theory: the limits of universalism, in H. Levin (ed.) *Community Control of Schools* (Washington, DC: Brookings Institution), 76–99.

FIRESTONE, W. (1990) Continuity and incrementalism after all: state responses in the excellence movement, in J. Murphy (ed.) *The Educational Reform Movement of the 1980s: Perspectives and Cases* (Berkeley, CA: McCutchan), 143–166.

FUHRMAN, S. and ROSENTHAL, A. (eds) (1981) *Shaping Education Policy in the States* (Washington, DC: Institute for Educational Leadership).

GARDNER, S. (1992) Key issues in developing school-linked, integrated services, *The Future of Children*, 3(1): 85–95.

GARMS, W., GUTHRIE, J. and PIERCE, L. (1978) *School Finance: The Economics and Politics of Education* (Englewood Cliffs, NJ: Prentice Hall).

GUTHRIE, J. (1988) Educational finance: the lower schools, in N. Boyan (ed.) *Handbook of Research on Educational Administration* (New York: Longman), 373–390.

HASTINGS, A. (1980) *The Study of Politics and Education: A Bibliographic Guide to the Research Literature* (Eugene, OR: ERIC Clearinghouse on Educational Management).

HAWLEY, W. (1985) False premises, false promises: the mythical character of public discussion about education, *Phi Delta Kappan*, November: 183–187.

HEINEKE, W. and BRAND, W. (1994) Assumptive worlds of local school boards: a case study, a paper presented at the Annual Meeting of the American Educational Research Association, New Orleans, LA.

IANNACCONE, L. (1975) *Educational Policy Systems* (Ft. Lauderdale, FL: Nova University Press).

IANNACCONE, L. (1977) Three views of change in educational politics, in J. Scribner (ed.) *The Politics of Education: The Seventy-sixth Yearbook of the National Society for the Study of Education* (Chicago: The University of Chicago Press), 255–286.

IANNACCONE, L. (1985) Excellence: an emergent educational issue, *Politics of Education Bulletin*, 12(3): 1, 3–8, 12.

IANNACCONE, L. (1988) From equity to excellence: political context and dynamics, in W. Boyd and C. Kerchner (eds) *The Politics of Excellence and Choice in Education: The 1987 Yearbook of the Politics of Education Association* (New York: Falmer Press), 49–66.

IANNACCONE, L. (1991) Micropolitics of education: what and why, *Education and Urban Society*, 23(4): 465–471.

IANNACCONE, L. and LUTZ, F. (1970) *Politics, Power, and Policy: The Governing of Local School Districts* (Columbus: OH: Merrill).

JEHL, J. and KIRST, M. (1992) Getting ready to provide school-linked services: what schools must do, *The Future of Children*, 2(1): 95–106.

JENNINGS, J. (1992) Lessons learned in Washington, DC, *Phi Delta Kappan*, 7: 303–307.

JORDAN, K. and LYONS, T. (1992) *Financing Public Education in an Era of Change* (Bloomington, IN: Phi Delta Kappa Educational Foundation).

KAUFMAN, H. (1956) Emerging conflicts in the doctrines of public administration, *American Political Science Review*, 50(4): 1057–1073.

KIMBROUGH, R. (1964) *Political Power and Educational Decision-Making* (Chicago: Rand McNally).

KIRBY, D., HARRIS, T. and CRAIN, R. (1973) *Political Strategies in Northern School Desegregation* (Lexington, MA: D. C. Heath).

LASSWELL, H. (1936) *Politics: Who Gets What, When, How* (New York: McGraw-Hill).

LEVIN, H. (ed.) (1970) *Community Control of Schools* (Washington, DC: Brookings Institution).

LUTZ, F. (1984) The people, their politics, and their schools, *Issues in Education*, II(2): 136–145.

LUTZ, F. and GRESSON, A., III (1980) Local school boards as political councils, *Educational Studies*, 11(2): 125–144.

LUTZ, F. and IANNACCONE, L. (eds) (1978) *Public Participation in Local School Districts* (Lexington, MA: D. C. Heath).

MALEN, B. and OGAWA, R. (1988) Professional-patron influence on site-based governance councils: a confounding case, *Educational Evaluation and Policy Analysis*, 10(4): 251–170.

MARSHALL, C. and SCRIBNER, J. (guest eds) (1991a) *Education and Urban Society*, 23(4).

MARSHALL, C. and SCRIBNER, J. (1991b) It's all political: inquiry into the micropolitics of education, *Education and Urban Society*, 23(4): 347–355.

MARSHALL, C., MITCHELL, D. and WIRT, F. (1989) *Culture and Education Policy in the American States* (New York: Falmer Press).

MAZZONI, T. (1993) The changing politics of state education policy making: a 20-year Minnesota perspective, *Educational Evaluation and Policy Analysis*, 15(4): 357–380.

MELAVILLE, A. and BLANK, M. (1993) *Together We Can* (Washington, DC: US Department of Education).

MITCHELL, D. (1974) Ideology and public school policy-making, *Urban Education*, 7: 35–59.

MITCHELL, D. (1980) The ideological factor in school politics, *Education and Urban Society*, 12: 436–451.

MITCHELL, D. (1988) Educational politics and policy: the state level, in N. Boyan (ed.) *Handbook of Research on Educational Administration* (New York: Longman), 453–462.

NATIONAL EDUCATION ASSOCIATION (1984) *Report of the Committee of Ten on Secondary School Studies; With the Reports of the Conferences Arranged by the Committee* (New York: published for the National Education Association by the American Book Company).

OAKES, J. (1985) *Keeping Track: How Schools Structure Inequality* (New Haven, CT: Yale University Press).

O'REILLY, J. (1988) Effects of principals' scheduling allocations on teachers' instructional strategies, unpublished doctoral dissertation, Arizona State University.

PAGE, A. and CLELLAND, D. (1978) The Kanawha County textbook controversy: a study of the politics of life style concern, *Social Forces*, 57: 265–281.

PESHKIN, A. (1978) *Growing up American* (Chicago: University of Chicago Press).

POWELL, A., FARRAR, E. and COHEN, D. (1985) *The Shopping Mall High School: Winners and Losers in the Academic Marketplace* (Boston: Houghton Mifflin).

PROSEMINAR ON EDUCATION POLICY (1991) *New Arizona Schools: An Analysis of the Governor's Task Force on Educational Reform* (Tempe, AZ: Arizona State University Education Policy Studies Laboratory).

RAY, C. and MICKELSON, R. (1990) Business leaders and the politics of school reform, in D. Mitchell and M. Goertz (eds) *Education Politics for the New Century: The Twentieth Anniversary Yearbook of the Politics of Education Association* (London: Falmer Press), 119–134.

Reaching the Goals (a series of working papers by various National Education Goals Work Groups, OERI; Washington, DC: US Department of Education).

REYES, P. and CAPPER, C. (1991) Urban principals: a critical perspective on the context of minority student dropout, *Educational Administration Quarterly*, 27(4): 530–557.

RODRIGUEZ V. SAN ANTONIO, 411 U.S. 1 (1973).

SCHWAGER, M., MITCHELL, D., MITCHELL, T., and HECHT, J. (1992) How school district policy influences grade level retention in elementary schools, *Educational Evaluation and Policy Analysis*, 14(4): 421–438.

SCRIBNER, J. and ENGLERT, R. (1977) The politics of education: an introduction, in J. Scribner (ed.) *The Politics of Education: The Seventy-sixth Yearbook of The National Society for the Study of Education* (Chicago: University of Chicago Press), 1–29.

SPRING, J. (1988) *Conflict of Interests: The Politics of American Education* (New York: Longman).

STOUT, J. (1982) *The enculturation of new school board members: a longitudinal study in seven school districts*, unpublished doctoral dissertation, Arizona State University.

STOUT, R. (1986) State reform as blitz, *Peabody Journal of Education*, 63(4): 100–106.

STOUT, R. (1993) Enhancement of public education for excellence, *Education and Urban Society*, 25(3): 300–310.

TALLERICO, M. (1989) The dynamics of superintendent–school board relationships: a continuing challenge, *Urban Education*, 24(2): 215–232.

TIMAR, T. (1989) The politics of school restructuring, *Phi Delta Kappan*, December: 265–275.

US GENERAL ACCOUNTING OFFICE (1993) *Educational Achievement Standards: NAGB's Approach Yields Misleading Interpretations* (Washington, DC: US General Accounting Office).

WENINGER, T. and STOUT, R. (1989) Dissatisfaction theory: policy change as a function of school board member–superintendent turnover, *Educational Administration Quarterly*, 25(2): 162–180.

WHITT, E., CLARK, D. and ASTUTO, T. (1986) *An Analysis of Public Support for the Educational Policy Preferences of the Reagan Administration*, Occasional Paper No. 3 of the Policy Studies Center of the University Council for Educational Administration (Charlottesville, VA: Curry School of Education, University of Virginia).

WILLIE, C. (1984) *School Desegregation Plans That Work* (Westport, CT: Greenwood Press).

WIRT, F. and KIRST, M. (1992) *Schools in Conflict*, 3rd edn (Berkeley, CA: McCutchan).

WOLCOTT, H. (1977) *Teachers Versus Technocrats: An Educational Innovation in Anthropological Perspective* (Eugene, OR: University of Oregon Center for Educational Policy and Management).

2.   The politics of education: from political science to multi-disciplinary inquiry

Kenneth K. Wong

The politics of education as a field of study owes much of its intellectual roots to political science. Its primary concerns clearly fall in the domain of political science – power, influence, conflict, and the 'authoritative allocation of values' (Easton 1965, Peterson 1974, Layton 1982, Burlingame 1988). This close relationship, however, is neither uni-directional nor static. While heavily relying on the methods and concepts in political science, the politics of education field has in turn contributed to theory building in the 'parent' discipline. To a large extent, then, the close interaction has been productive and mutually rewarding.

Close connection to the parent discipline notwithstanding, the field of education politics has vigorously adopted concepts and methods from other disciplines. The divergence from political science becomes most pronounced during the 1980s, a period when economic science gained prominence among political scientists but not among the politics of education researchers. In my view, the divergence poses both a blessing and a challenge. On the one hand, it is encouraging to see researchers in the field of educational politics draw on multidisciplinary perspectives to produce a knowledge base that is useful to both the academic and the policy communities. In doing so, the field progresses toward defining its own intellectual identity. On the other hand, the field's broadening research strategies create an intellectual challenge. At issue is the degree of coherence among politics of education researchers. What are the common conceptual threads that weave the field together as its members approach school politics with diverse analytical lenses? This chapter will address these issues that confront our profession as the Politics of Education Association moves into the second 25 years.

Overall, the transformation in politics of education study toward a broadened identity must be seen as remarkable given the field's relatively young age as an organized profession. This chapter examines this transformation in the context of the changing relationship between the field and the political science discipline. It should be mentioned that this review chapter cannot address all major topics in the field. Limited by my own expertise, I will not review the scholarship in international and comparative education. Nor will I examine the literature on political socialization.

The following chapter is organized into four sections. First, I discuss the extent to which political science has shaped the politics of education field. Then, the field's contribution to theory-building in political science will be examined. Third, I explore several reasons why politics of education researchers diverge from the new paradigm in political science, namely the rational choice model. The final section discusses the implications of the field's multidisciplinary character for future research strategies.

0268–0939/94 $10·00 © 1994 Taylor & Francis Ltd.

When political science reigns

It comes as no surprise that much of the work in the field of the politics of education is grounded in the basic subject matter in political science. In his seminal *American Political Science Review* article that preceded the organization of the Politics of Education Association, Thomas Eliot (1959) developed a set of research strategies for the 'continuing analysis of how the schools are run and who runs them' (p. 1932). Studies at the district level, according to Eliot, would include 'structural analysis' on the ways districts are organized, 'behavioral analysis' that examines the impact of professional and citizen leadership and interest group activities on policy decisions, and voting analysis on elections and bond issues. Eliot also suggested the need to examine state and federal educational policy, finance, and politics.

Clearly, Eliot's proposed research agenda has been taken seriously by students in the politics of education. Almost 20 years later, in their major review of the development of the field, Scribner and Englert (1977) concluded that the robust scholarship in the politics of education has as its primary boundary the inquiry into interactions that influence the 'authoritative allocation of values' (see Easton 1965). In both substantive and analytical terms, the field has pursued a line of research that is quite consistent with what Eliot proposed earlier. By the 1970s, according to Scribner and Englert, much of the work in the field focused on four issues – the way educational governance is organized, the distribution of power, the nature and management of conflict, and the outcomes and impact of educational policies.

Indeed, these concerns about the political functions of various collective entities have produced a voluminous amount of research during the first 25 years of PEA as indicated in the literature reviews by, among others, Mazzoni (1993) and Mitchell (1988) on state policy-making and interest groups, Burlingame (1988) and Wong (1992) on local school politics, LaNoue (1982), Layton (1982), and Wirt and Kirst (1982) on institutions, political culture, and systems analysis, Zeigler, Jennings and Peak (1974) on decision-making and power, and Boyd (1983) and Fuhrman (1993) on policy design and content.

Because the PEA was founded at a time when 'behavioralism' dominated political science, it is not surprising that many politics of education scholars are in part influenced by this approach. Behavioralism, as Robert Dahl pointed out, 'is an attempt to improve our understanding of politics by seeking to explain the empirical aspects of political life by means of methods, theories, and criteria of proof that are acceptable according to the canons, conventions, and assumptions of modern empirical science' (1969: 77). It involves the construction of a set of testable hypotheses to guide and organize empirical evidence (see Zeigler and Jennings 1974). At the same time, politics of education scholarship tends to avoid two limitations of the 'behavioralists' (who focus on what is), namely a failure to appreciate history (what has been) (see Peterson 1985) and the lack of normative discourse (what ought to be) (Hochschild 1984, Gutmann 1987).

Overall, politics of education researchers are ready to use political science methods and approaches because, for the most part, they share the view that school governance and decisions are embedded in the core practices of our political systems. Educational policy, as Paul Peterson (1974) pointed out, routinely features a low level of citizen participation, a central board that serves as an agent of legitimization, and considerable influence from a small group of professional staff. However, the mayor and other public officials are expected to be actively involved in school (just as in other policy) affairs when 'the issue has major budgetary implications, generates a widespread community controversy, or involves the jurisdiction of other local agencies' (1974: 365). In evaluating the key contribution of the field, Douglas Mitchell observed, 'We now recognize that both the content and the form of schooling is determined through the conflicts and coalitions found at the core of local, state and national political systems' (1990: 166). In short, politics of education research has clearly suggested the ways in which education policy is shaped by the broader political institutions and decision-making processes.

Contributions of politics of education study to political science

More importantly, in establishing the linkage between politics and schools, the politics of education scholarship not only benefits from the 'parent' discipline, but also contributes to the development and refinement of conceptual frameworks in mainstream political science. In some cases, politics of education research informs policy decisions as well. Let me cite a few examples.

Federalism revisited

The politics of education scholarship has helped revise our contemporary understandings of administrative cooperation and conflict in intergovernmental relations. During the 1960s, the theory of federalism was dominated by Morton Grodzins's (1966) metaphor of the 'marble cake,' where intergovernmental relations are marked by mutual assistance rather than confrontation between layers of the government. Consistent with this view, Great Society programs such as compensatory education were financed by the federal government but implemented by the state and local agencies. As knowledge accumulated from lessons of program implementation, the 'marble cake' concept was subject to empirical challenge. Among the seminal works that found a great deal of intergovernmental conflict in the provision of federally funded programs was Jerome Murphy's (1971) detailed analysis of the ESEA Title I program in Massachussetts. The study documented extensive local noncompliance with the federal intent of targeting services to disadvantaged students. Murphy attributed the program failure to the lack of organized 'political pressure by poverty groups and their allies, and to meaningful participation by subinterest in local school district councils or other comparable devices.'

Findings like these and others (Derthick 1972, Pressman and Wildavsky 1973, Coleman et al. 1975), together with policy reports that came from the NAACP Children's Defense Fund (Martin and McClure 1969) and other organizations, provided the basis for greater federal efforts to monitor local use of program funds. More importantly, analyses of intergovernmental conflicts have revised the way political scientists view federalism. Instead of the 'marble cake,' the theory of federalism during the 1970s and the early 1980s was dominated by the 'implementation' literature. The latter saw organizational confusion, institutional conflict, and bureaucratic red tape in the federal grant system. The organizational structure, according to some, can be characterized as 'picket fence federalism' (Wright 1982). Based in part on the implementation view, the Nixon, Ford, Reagan, and Bush administrations called for a decentralized federal grant system.

By the mid- and late-1980s, the implementation view of federal–local conflict was criticized by Peterson, Rabe and Wong (1986; also see Peterson 1981, Wong 1990). Based on a comparative analysis of federal policy in education, health, and housing, the authors presented a 'differentiated theory of federalism.' First, the study distinguishes two patterns of intergovernmental relations that are closely related to the purpose of federal policy. While conflict occurs in redistributive policy, intergovernmental cooperation remains strong in developmental programs. Second, with the passage of time, even redistributive programs become more manageable. With a few exceptions, local administrative agencies were found to be in compliance with the federal targeting intent. These scholarly efforts to sort out and specify the functions of the federal and the local governments tend to provide the empirical basis for maintaining the redistributive character of many categorical social programs. Consequently, during the midst of the Reagan retrenchment in fiscal year 1984, over 80% of the federal grants in education, training, and employment-related services went to special-needs groups (Wong 1989).

With the arrival of the Clinton administration, the role of the federal government is again subject to close examination. Policy coherence tops the agenda (Smith and O'Day 1991, Fuhrman 1993, O'Day and Smith 1993). In this context of systemic restructuring, Martin Orland (1993) proposed to replace 'picket fence federalism' with 'chain link federalism.' While the former sustains organizational fragmentation, the latter is designed to foster interdependence among various policy components. Orland sees the importance of placing disadvantaged students within the context of broader systemic reform, where policy is designed to facilitate problem-solving skills for all students, emphasize outcome-based accountability, and maintain organizational coherence. Clearly, continual rethinking among politics of education researchers about the ways federalism works in education is critical to theory rebuilding in the political science discipline.

Going beyond the 'elitist' versus 'pluralist' debate

The politics of education scholarship, given its strong focus on local school governance, has advanced our conceptualization of power structure and democratic practices. Because school governance is rooted in our beliefs in democratic control, politics of education scholarship has provided valuable information that enriches the debate among political scientists over power structure. To be sure, there is significant disagreement over the extent to which power and influence is widely distributed. On the one hand, the 'élitist' framework centers on how the community's economic and political élites closely dominate school policy making (Hunter 1953, Bachrach and Baratz 1962, McCarty and Ramsey 1971, Stone 1976, Spring 1988).

Indeed, even by the late 1960s and early 1970s, Zeigler and Jennings (1974) found that public school governance was largely closed to lay influence, a finding that contradicts democratic norms. On the other hand, the 'pluralist' perspective focuses on the extent to which school policies are influenced by competing interests, partisan contention, and electoral constraint (Dahl 1961, Grimshaw 1979, McDonnell and Pascal 1979, Wirt and Kirst 1982). Iannaccone (1967) proposed four types of interest group politics that capture the development of relationships between competing interests and state lawmakers. Various case studies have found that Iannaccone's Type III model (statewide fragmented) best describes the current power structure (Nystrand 1976, Kirst and Somers 1981, also see Marshall et al. 1989, McGivney 1984, Mazzoni 1993).

Building on the élitist-pluralist tradition, politics of education scholarship has advanced our understanding of the tension between the authority structure and democratic norms. Three bodies of literature provide useful examples – (1) multiple centers of power, (2) race relations, and (3) democracy and education.

Autonomous centers of power: One set of studies is concerned about the emergence of seemingly autonomous power centers in shaping school policy. One contending interest is an increasingly skeptical property taxpayer population. In many cities, a substantial number of middle-class families no longer enrolls their children in public schools (Kirst and Garms 1980, Wong forthcoming). The aging of the city's taxpaying population has placed public education in competition with transportation, hospital, and community development over local tax revenues.

Another center of power is the teachers' union. Grimshaw's (1979) historical study of Chicago's teachers' union suggests that the union has gone through two phases in its relationship with the city's political machine and school administration. During the formative years, the union largely cooperated with the political machine in return for a legitimate role in the policy-making process. In the second phase, with Grimshaw characterized as 'union rule,' the union became independent of either the local political machine or the reform factions. Instead, it looked to the national union leadership for guidance and engaged in tough bargaining with the administration over better compensation and working conditions. Consequently, Grimshaw argued that policy makers 'no longer are able to set policy unless the policy is consistent with the union's objectives' (1979: 150).

In the current reform climate, greater attention has been given to rethinking the role of the union. The Chicago Teachers' Union has established the Quest Center to strengthen professional development (see Ayers 1993). Drawing on lessons from various districts where unions have become change agents, Kerchner and Caufman (1993) proposed a framework for 'professional unionism.' This new organizing system has three characteristics. First, traditional separation between management and labor will be replaced by a mode of shared operation (e.g., team teaching, site decision making). Second, adversarial relationships will give way to a strong sense of professional commitment and dedication. Third, unions will incorporate 'the larger interests of teaching as an occupation and education as an institution' (p. 19). These emerging concepts are likely to revise our seemingly dated understanding of labor union politics, which to a large extent is based on the industrial relations model developed during the New Deal of the 1930s and the 1940s.

Managing race relations: The politics of education scholarship has also contributed to our understanding of governmental efforts to promote racial equity (Orfield 1969, Hawley *et al*. 1983, Hochschild 1984, Rossell 1990). Increases in minority representation in recent years have contributed to an improvement in educational equity. In the post-*Brown* and post-Civil Rights Movement era, big-city districts began to respond to demands from minority groups by decentralizing certain decision-making powers to the community (see Reed 1992). By the 1980s, parent empowerment at school sites had gained support from reform interest groups, businesses, and elected officials. Minority groups have also gained representation at the district level of leadership. In the 1980s, many major urban districts were governed by a school board dominated by minority representatives, who, in turn, selected a minority individual to fill the position of school superintendent (Jackson and Cibulka 1992).

Minority representation has improved educational equity for minority students in at least two ways. First, minority groups can put pressure on the school bureaucracy to allocate resources in an equitable manner (Rogers 1968, Ravitch 1974). As recently as 1986, a coalition of black and Hispanic groups filed suit against the Los Angeles school district for failing to provide equal resources and experienced teachers to predominantly minority schools in the inner-city neighborhoods (*Rodriguez* v. *Los Angeles Unified School*

District). Five years later, the litigation was brought to an end when the state superior court approved a consent decree requiring the district to equalize the distribution of experienced teachers among schools and to allocate basic resources and supplies on an equal, per-pupil basis (*Education Week* 9 September 1992). Similar organized actions are found in other cities where the school population has undergone major demographic changes.

Further, minority representation affects personnel policy, which in turn may have instructional consequences for disadvantaged pupils. Using data from the Office of Civil Rights in districts with at least 15,000 students and 1% black, Meier *et al.* (1989) examined the practice of second-generation discrimination in the classroom following the implementation of school desegregation plans. They found that black representation on the school board has contributed to the recruitment of black administrators, who in turn have hired more black teachers. Black teachers, according to this study, are crucial in reducing the assignment of black students to classes for the educable mentally retarded. Black representation in the instructional staff also reduces the number of disciplinary actions against black students and increases the latter's participation in classes for the gifted. Another study found that increases in the number of Hispanic teachers tends to reduce dropout rates and increase college attendance for Hispanic students (Fraga *et al.* 1986). In other words, minority representation tends to reduce discriminatory practices and facilitate equal opportunities in the classroom.

Rethinking democracy and education: Concerns over participatory democracy have also renewed our interest in the ways in which formal schooling contributes to our civic community (Dewey 1916, also see Barber 1994). In addressing the question of the moral boundary of formal schooling, political theorist Amy Gutmann (1987) offers a framework that is based on principled limits on democratic governance. The two key principles are 'nonrepression' that allows for competing ideas and 'nondiscrimination' that does not exclude anyone from democratic participation and education. To achieve these democratic goals, she sees the need for collective responsibilities from the state (Plato's notion of 'Family State'), parents (Locke's concept of 'State of Families'), and professional educators (Mill's idea of 'State of Individuals'). Equally concerned about the extent to which social structure undermines democratic schooling, Katznelson and Weir (1985) argue that the United States no longer practices the ideal of 'common schooling' where children of diverse class backgrounds are expected to obtain similar learning experiences in their neighborhood schools across districts. Through a historical analysis of three urban districts, Katznelson and Weir conclude that the coalition in support of egalitarian public education has been compromised both by the narrowing definition of schooling, primarily in terms of instruction and curriculum, and the declining political influence of the labor movement.

Consequently, educational inequity is embedded in today's social geography, most notably the isolation of central cities. These concerns about the broader societal role of education are also shared by scholars overseas. For example, in November 1991, the University of Chicago organized a conference on 'Democracy and Education' with invited speakers from the former Soviet Bloc countries. The transition from a state-directed into a market-oriented economy clearly has significant implications on how schools should be organized and what should be taught in the post-Soviet era.

Divergence from political science

By the 1980s it was clear that political science was undergoing a significant transformation. The process-oriented tradition established by the works of Bentley, Truman, and Dahl has been pushed aside by the rational choice model. Case studies on the process of decision making are replaced by econometric analysis and mathematical languages. Institutional tradition (e.g., party affiliation), organizational culture (e.g., centralization and division of labor), and collective entities (e.g., social movement, interest groups, legislature) are increasingly seen as incomplete in understanding public decisions.

Instead, a new paradigm of rational choice that uses individuals as the theoretical unit of analysis has gained prominence in political science. As Peter Ordeshook (1986) points out, '[A]ny adequate understanding of group choice or action ultimately must be reducible to an understanding of the choices that individual human beings make in the context of institutions for the purpose of attaining individual objectives' (p. xii). In synthesizing economics and politics, a new generation of political scientists vigorously apply the rational choice model to voting behavior, bureaucratic decisions, and legislative actions (see Downs 1957, Black 1958, Buchanan and Tullock 1962, Riker 1962, and Shepsle 1989).

Deeply concerned with the blurring of the disciplinary boundary between economics and political science, Theodore Lowi, in his 1991 presidential address to the American Political Science Association, warns his peers in the profession against losing their own intellectual identity (1992). In his review over the development of the profession, Lowi observed that its founding generation (around 1890s through 1920s) grew up during the progressive movement and treated politics as a problem for representative government. At that time, political scientists were committed to distinguishing between the formal, legal system of democratic governance and the realities of political power and influence. As Woodrown Wilson in his presidential address before the APSA in 1911 stated, 'I take the science of politics to be the accurate and detailed observation of [the] process by which the lessons of experience are brought into the field of consciousness, transmuted into active purposes, put under the scrutiny of discussion, sifted, and at least given determinate form in law' (as cited in Lowi 1992: 1).

As the national government expanded its scope beginning with the New Deal during the 1930s, the increasingly bureaucratized state has become 'intensely committed to science.' According to Lowi, science not only influences the ways policy decisions are made, it also reshapes the ways political scientists define their work. Among the sciences, it is economics that 'replaced law as the language of the state' (Lowi 1992: 3). Needless to say, the symbiotic relationship between economics and policy making has altered the discipline of political science. The discipline's two prominent subfields, public policy analysis and public choice, have largely replaced traditional public administration and institutional analysis. Both subfields now rely heavily on economic thinking and methods. Building on the earlier works of Herbert Simon and others, researchers in the two subfields use rationality at the lowest decision-making unit (i.e., individuals) to explain and predict collective behaviors and policy choices. For example, political scientists use game theory (e.g., prisoners' dilemma) to explain such wide-ranging subjects as diplomacy and interest group politics.

From Lowi's perspective, the marriage of economics and political science has exacted great costs to the profession. First, 'Policymaking powers are delegated less to the agency and more to the decisionmaking formulas residing in the agency' (1992: 5). Second, the profession has not systematically examined the political functions of economic tools in policy making. For example, economic tools (e.g., use of indices) may temper policy deliberation and may trivialize citizen participation. Finally, economic analysis of political problems has made the profession a 'dismal science,' resulting in the loss of 'passion.' The latter, according to Lowi (1992: 6) includes 'the pleasure of finding a pattern, the inspiration of a well-rounded argument, the satisfaction in having made a good guess about what makes democracy work and a good stab at improving the prospect of rationality in human behavior.' Very rarely do we now find political scientists who are willing to invest their time carefully observing the daily operation of an institution at close range in the tradition of Sundquist (presidential–congressional relations), Dahl, Banfield, Greenstone, Polsby, and Wolfinger (urban politics), Peterson (school politics in Chicago), Wilson (bureaucracy and interest groups), Fenno (Congress), Orfield (national policy on school desegregation), and Neustadt (presidency).

Instead of following the path taken by political science, the politics of education field seems reluctant to embrace economic science. In this regard, the politics of education field departs from its 'parent' discipline. The divergence became apparent by the 1980s, as illuminated by the debate over school choice. In an attempt explicitly to link politics and bureaucracy to school effectiveness, two political scientists, John Chubb and Terry Moe (1990), examined the relationship between school autonomy (as opposed to direct electoral and bureaucratic control) and student performance in high schools. In their view, educational governance is an open system where political interests have successfully expanded the bureaucracy and proliferated programmatic rules to protect their gains. Their perspectives is rooted in the earlier work by economists, including Milton Friedman (1962) who argued for market competition and parental choices in education. Chubb and Moe's analysis tends to support Friedman's position. Using the High School and Beyond surveys for 1982 and 1984 and the Administrator and Teacher Survey data for 1984, Chubb and Moe found that politics and bureaucracy do not contribute to the desirable forms of school organization that link to higher academic performance of their students. Instead, the more market-oriented nonpublic schools are far more likely to produce what they call effective organizations.

No doubt, the Chubb and Moe study has received a great deal of attention and has prompted politics of education researchers to examine school choice and other mechanisms to promote efficiency (Boyd and Walberg 1990, Weeres and Cooper 1992). However, most politics of education researchers are hesitant to endorse school choice. Michael Kirst (1990), for example, is critical of Chubb and Moe's heavy reliance on 116 items from five sets of tests that took 63 minutes to complete. David Kirp (1990) was skeptical about any reform that is primarily based on the concept of efficiency (particularly when it is narrowly defined by Chubb and Moe as improvement in achievement test scores between 10th and 12th grades). From Kirp's perspective, 'Politics has its place in shaping a public education . . . [in developing a] sense of schools as belonging to, defining, and being defined by, a community' (1990: 41). Likewise, Richard Elmore (1991) raises questions about equality under the proposed choice system. In his view, the analysis 'simply legitimates sorting students among schools by race, social class, and the capacity of their parents to manipulate the ground rules of choice to their own advantage' (p. 693). Elmore concludes that school reform is much more complex than Chubb and Moe suggest. Critical to thinking about reform, but largely absent from the analysis, are teacher policy, student diversity, and instructional organization (Elmore 1991: 694).

A closer look at the politics of education field itself suggests several reasons why politics of education researchers have not embraced economic thinking in political and policy analyses. First, the field, though primarily associated with developments in political science, has adopted perspectives from a variety of disciplines, including, among others, sociology of education (see Barr and Dreeben 1983, Bidwell 1992), anthropology (Hess 1992), public administration (Guthrie 1990), and human development and curriculum theories (see Fuhrman and Malen 1991).

Of equal importance, scholarship in the politics of education remains practice-oriented and can be broadly characterized as applied research. Although some draw on national databases (e.g., High School and Beyond and the Principals and Administrators Surveys), many researchers continue to gather information directly from the schools, communities, districts, states, and the federal government. Conversations with the policy community are frequent and efforts are made to disseminate research findings to practitioners. Efforts to link research to practitioners are facilitated by the federally funded Consortium for Policy Research in Education as well as over two dozen state- and foundation-funded policy centers. This latter practice is in sharp contrast with political science, where the leading journals in recent years have been filled with econometric models and mathematical formulae such that only the most well-trained quantitative analysts can appreciate their policy implications.

Further, politics of education researchers are actively engaged in not only problem identification but also problem solving in areas that confront policy makers and practitioners. The politics of education yearbooks, for example, have addressed such important policy issues as urban education, school desegregation, assessment and curriculum, service integration, race and gender equity issues, and school finance. These are clearly issues that are central to the functioning and improvement of the public education system.

The expanding, multidisciplinary scope in politics of education scholarship is indicative of a field that is growing in membership, activities, and policy concerns. The field has a healthy mix of researchers whose training and interest range from educational administration, policy analysis, school finance, state and federal issues, private versus public schools, and organization at the school and classroom level. Some politics of education researchers have extensive administrative experience in the educational sector and share their knowledge on how schools and school systems actually operate. In many universities and colleges, the field of politics of education has merged with administration and policy programs. Its curriculum is often closely related to the broader mission of school governance and policy improvement.

Toward defining the field's own identity: implications for future research

Given its multidisciplinary approach and its connection to practice, where is the politics of education field heading? The challenge for the field is whether it constitutes a 'synthetic entity' as it moves from primarily political science-based inquiry to a multidisciplinary approach. Is the field little more than a collection of studies that are situational yet loosely connected to each other? Do researchers see their works as falling within a coherent theoretical construct that defines the ways school politics occur? Is it possible or realistic to suggest elements that tend to define the field's own identity? A useful starting point in thinking about these questions is to look at the extent to which the field has been redefined as a result of cross-fertilization.

Cross-fertilization has resulted in several changes in the ways we have studied educational politics over the past three decades. These changes may provide the analytical basis for our research agenda in the future. In this section, I will briefly explore several possibilities in research strategies that may help define our research direction, thereby shaping our organization's future outlook.

Public school reform remains our focus

Researchers in the politics of education are expected to engage in extensive examination of major efforts to improve public education. From a broad perspective, reform efforts over the past three decades tended to oscillate between two perspectives that can be roughly placed along an 'equity–efficiency' continuum. While the 1960s and the 1970s were years of equal opportunities, the 1980s redirected our attention to efficiency.

As we enter the mid-1990s, there are several emerging policy directions that provide research opportunities. The Clinton Administration has assumed greater responsibility for improving governmental performance. Like Reagan, Clinton focuses on productivity in a period of budgetary constraint. Unlike Reagan, Clinton clarifies the federal role. Instead of deferring responsibility to states and market-like forces, the Clinton Administration specifies areas where federal leadership should be exerted and areas where a partnership between the federal government and other entities would be constructive.

The current reform climate provides an excellent opportunity for vigorous research on the politics of national educational policy (see the chapter by Sroufe in this volume). The Clinton Administration has undertaken several initiatives to enhance policy coherence and accountability. First, seeing a linkage between pre-school services and instructional programs in the lower grades, the Administration strongly supports early childhood intervention strategies. Second, federal leadership is directed at setting new national standards and designing a national examination system in five core areas – English, mathematics, social studies, science, and foreign languages. Third, new standards are considered for teachers and other professions as well as for the school organization itself (the latter is broadly labeled as School Delivery Standards).

Further, federal efforts are made to address organizational fragmentation and inter-governmental contention. One major attempt to streamline the federal bureaucracy is outlined by the 1993 report conducted by the Commission on National Performance Review, chaired by Vice-President Al Gore. The report called for trimming the overall federal workforce by 12%. It suggests the elimination of almost one-sixth of the 230 programs that are administered under the Department of Education. To reduce duplication, the Gore report recommends extensive program consolidation both within the education agency and between education and other agencies (such as the Labor Department). It also recommends the creation of flexible block grants for state agencies.

Another effort toward greater accountability is the redesigning of the federal compensatory education program (Chapter 1). The Administration-proposed Improving America's School Act of 1993 is directed at the overall quality of the schools that poor children attend (HR3130). It plans to reallocate Chapter 1 dollars from the more affluent districts to poor districts, raise the standards for low-achieving students, and encourage instruction and curriculum that teach students critical thinking skills. The Administration's bill is clearly consistent with major assessments of Chapter 1. As the 1992 report of the Commission on Chapter 1 argues, Chapter 1 can be substantially improved with a new accountability framework that aims at 'producing good schools, not simply good programs.' Instead of mandating schools to meet accounting standards, the Commission recommends that schools be accountable for student progress in learning.

Congress is expected to adopt most of these new ideas, although it has rejected the administration's proposal to channel more money to high-poverty schools. Overall, the current restructuring would reduce the 'categorical' nature of Chapter 1, thereby enhancing programmatic coordination between federal, state, and local staff in poor schools. If Chapter 1 is redesigned in ways that would produce better schools in poor neighborhoods, then the Clinton Administration can rightfully claim that equity and efficiency can be pursued simultaneously.

Given the enduring and complex relationship between equity and efficiency, politics of education analysts are expected to conduct systematic study on these issues as they evolve from the legislative to the implementation phase. New initiatives may be launched and existing practices may be altered owing to changing circumstances. The controversy over choice in the name of efficiency throughout the 1980s and the early 1990s is a good example of how policy analysts gravitate to a new set of issues. The debate not only polarizes the choice proponents and public-sector supporters but also enhances the position of those who call for choice within public schools.

Equally important is the research on the ways in which existing concepts are altered under different circumstances. The notions of decentralization and empowerment are good examples. Whereas decentralization has often been associated with New York style community control to promote racial representation since the 1960s, the concept no longer denotes a common meaning as Chicago advances its own ambitious version of parent empowerment and Dade County and other districts implement site-based management. Further, when one expands the service delivery system to include higher levels of government, one may see an emerging trend of policy centralization both at the federal level (such as setting standards in content areas) and at the state level (such as creating a framework on student assessment).

Understanding service delivery in a multilayered organization

Politics of education research will continue to focus on teaching and learning in the multilayered policy organization. Given the complex organizational setting in education, we need to reconceptualize how schools work – the ways resources are used at different levels of the school organization, the organization and practice of teaching, and the grouping and distribution of curricular materials to students. In this regard, politics of education researchers have showed the macro–micro linkage between the policy-making level and the classrooms. Assessment of federal program implementation clearly suggests the importance of site-level variables (McLaughlin and Berman 1978, Elmore 1980, McLaughlin 1987).

Multidisciplinary inquiry

Analysts in the politics of education are confronted with a methodological task. As policy makers look for coherence in school services and formulate more comprehensive solutions to address chronic social problems, policy analysts have to become increasingly inter-disciplinary in conducting their research. Clearly, the politics of education field has been strengthened by adopting perspectives and tools from various disciplines. We have applied the concepts of human capital investment, incentives, and rational expectation from economics. From sociology, we learn about the nature and functions of bureaucracy, school organization, the process of producing learners, social capital, and the urban underclass. We see the importance of contextualizing our findings, as historians do. Like political scientists, we pay attention to the governance structure, the political process, interest groups, and the distribution of power. Taken as a whole, as discussed above, politics of education research has gone from a single discipline-based field to one that encompasses multiple orientations. The multidisciplinary approach will continue to define politics of education activities in the future.

Linking research to policy and practice

Ongoing reform efforts raise questions about whether research-based knowledge can be translated into policy and practice. Not surprisingly, there are political and organizational barriers, among other factors, to the dissemination and adoption of innovative practices in the public schools. The centralized bureaucracy has rarely transmitted university-based research to school-level personnel in an effective and timely manner (Cibulka 1992). The adherence to universalistic norms in resource allocation has also discouraged variation in local school practices. The incentive structure, which is not closely linked to student outcomes, does not create the necessary conditions for teachers to innovate. Despite these barriers, dissemination of policy research has been improved in recent years as government agencies and private foundations are more involved in the process of knowledge diffusion. In light of the communication gap between researchers and policy makers, politics of education scholarship over the past 25 years represents an effort to bridge the two communities. The next 25 years will continue to see a close linkage between politics of education research and the policy community.

To sum up, as the field becomes more diverse in approaches and methods, we will see more opportunities for collaborative undertakings. In the PEA's second 25 years, I believe there will be more joint initiatives between the academic and the policy communities as future reviews of our organization report on the impact of politics of education scholarship on school reform and policy improvement.

References

AYERS, W. (1993) Chicago: A restless sea of social forces, in C. Kerchner and J. Koppich (eds) *A Union of Professionals* (New York: Teachers College Press), Ch. 10.
BACHRACH, P. and BARATZ, M. (1962) The two faces of power, *American Political Science Review* (56): 947–952.
BARBER, B. (1994) *An Aristocracy of Everyone* (New York: Oxford University Press).
BARR, R. and DREEBEN, R. (1983) *How Schools Work* (Chicago: University of Chicago Press).
BIDWELL, C. E. (1992) Toward improved knowledge and policy on urban education, in J. Cibulka, R. Reed and K. Wong (eds) *The Politics of Urban Education in the United States* (London: Falmer Press), 193–199.
BLACK, D. (1958) *Theory of Committees and Elections* (Cambridge: Cambridge University Press).
BOYD, W. L. (1983) Rethinking educational policy and management: political science and educational administration in the 1980's, *American Journal of Education* (November): 1–29.
BOYD, W. L. and WALBERG, H. (1990) *Choice in Education: Potential and Problems* (Berkeley: McCutchan).
BUCHANAN, J. and TULLOCK, G. (1962) *The Calculus of Consent* (Ann Arbor: University of Michigan Press).
BURLINGAME, M. (1988) The politics of education and educational policy: the local level, in N. J. Boyan (ed.) *Handbook of Research on Educational Administration* (New York: Longman), 439–451.
CHUBB, J. and MOE, T. (1990) *Politics, Markets, and America's Schools* (Washington, DC: Brookings Institution).

CIBULKA, J. (1992) Urban education as a field of study: problems of knowledge and power, in J. Cibulka, R. Reed and K. Wong (eds) *The Politics of Urban Education in the United States* (London: Falmer Press), 27–43.

COLEMAN, J. S. *et al.* (1975) *Trends in School Desegregation 1968–73* (Washington, DC: Urban Institute).

DAHL, R. (1961) *Who Governs?* (New Haven: Yale University Press).

DAHL, R. (1969) The behavioral approach in political science: epitaph for a monument to a successful protest, in H. Eulau (ed.) *Behavioralism in Political Science* (New York: Atherton), Ch. 3, 68–92.

DERTHICK, M. (1972) *New Towns In Town: Why A Federal Program Failed* (Washington, DC: Urban Institute).

DEWEY, J. (1916) *Democracy and Education* (New York: Free Press).

DOWNS, A. (1957) *An Economic Theory of Democracy* (New York: Harper & Row).

EASTON, D. (1965) *A Systems Analysis of Political Life* (Chicago: University of Chicago Press).

ELIOT, T. (1959) Toward an understanding of public school politics, *American Political Science Review*, 53(4): 1032–1051.

ELMORE, R. (1991) Review of *Politics, Markets, and America's Schools, Journal of Policy Analysis and Management*, 10(4): 687–695.

ELMORE, R. (1980) Backward mapping: implementation research and policy decisions, *Political Science Quarterly*, 94(4): 601–616.

FRAGA, L. MEIER, K. and ENGLAND, R. (1986) Hispanic Americans and educational policy: Limits to equal access, *Journal of Politics*, 48: 850–876.

FRIEDMAN, M. (1962) *Capitalism and Freedom* (Chicago: University of Chicago Press).

FUHRMAN, S. (ed.) (1993) *Designing Coherent Education Policy* (San Francisco: Jossey-Bass).

FUHRMAN, S. and MALEN, B. (1991) *The Politics of Curriculum and Testing* (London: Falmer).

GRIMSHAW, W. (1979) *Union Rule in the Schools* (Lexington: Lexington Books).

GRODZINS, M. (1966) *The American System*, ed. D. Elazar (Chicago: Rand McNally).

GUTHRIE, J. (1990) The evolution of educational management: eroding myths and emerging models, in B. Mitchell and L. Cunningham (eds) *Educational Leadership and Changing Contexts of Families, Communities, and Schools*, Eighty-ninth Yearbook of the National Society for the Study of Education (Chicago: University of Chicago Press), Part II, 210–231.

GUTMANN, A. (1987) *Democratic Education* (Princeton: Princeton University Press).

HAWLEY, W. *et al.* (1983) *Strategies for Effective Desegregation* (Lexington, MA: Lexington Books).

HESS, G. A. (ed) (1992) *Empowering Teachers and Parents: School Restructuring Through the Eyes of Anthropologists* (New York: Bergin & Garvey).

HOCHSCHILD, J. (1984) *The New American Dilemma* (New Haven: Yale University Press).

HUNTER, F. (1953) *Community Power Structure* (Chapel Hill: University of North Carolina Press).

IANNACCONE, L. (1967) *Politics in Education* (New York: Center for Applied Research in Education).

JACKSON, B. and CIBULKA, J. (1992) Leadership turnover and business mobilization: the changing political ecology of urban school systems, in J. Cibulka, R. Reed and K. Wong (eds) *The Politics of Urban Education in the United States* (London: Falmer Press), 71–86.

KATZNELSON, I. and WEIR, M. (1985) *Schooling for All: Class, Race, and the Decline of the Democratic Ideal* (New York: Basic Books).

KERCHNER, C. and CAUFMAN, K. (1993) Guilding the airplane while it's rolling down the runway, in C. Kerchner and J. Koppich (eds) *A Union of Professionals* (New York: Teachers College Press), Ch. 1, 1–24.

KIRP, D. (1990) School choice is a panacea, these authors say, *The American School Board Journal*, 177(9): 38, 41.

KIRST, M. (1990) Review of *Politics, Markets, and America's Schools, Politics of Education Bulletin*, Fall.

KIRST, M. and GARMS, W. (1980) The political environment of school finance policy in the 1980's, in J. Guthrie (ed.) *School Finance Politics and Practices – The 1980s: A Decade of Conflict* (Cambridge: Ballinger), pp. 47–75.

KIRST, M. and SOMERS, S. (1981) California educational interest groups: collective action as a logical response to proposition 13, *Education and Urban Society* (13): 235–256.

LANOUE, G. R. (1982) Political science, in H. E. Mitzel *et al.* (eds) *Encyclopedia of Educational Research* (Washington, DC: American Educational Research Association), Vol. 3, 1421–1426.

LAYTON, D. (1982) The emergence of the politics of education as a field of study, in H. L. Gray (ed.) *The Management of Educational Institutions* (Lewes, UK: Falmer Press), 109–126.

LOWI, T. (1992) The state in political science: how we became what we study, *American Political Science Review*, 86(1): 1–7.

MARSHALL, C., MITCHELL, D. and WIRT, F. (1989) *Culture and Education Policy in the American States* (New York: Falmer Press).

MARTIN, R. and McCLURE, P. (1969) *Title I of ESEA: Is It Helping Poor Children?* (Washington, DC: Washington Research Project of the Southern Center for Studies in Public Policy and the NAACP Legal Defense of Education Fund).

MAZZONI, T. (1993) The changing politics of state education policy making: a 20-year Minnesota perspective, *Educational Evaluation and Policy Analysis*, 15(4): 357–379.

MACCARTY, D. and RAMSEY, C. (1971) *The School Managers: Power and Conflict in American Public Education* (Westport, CT: Greenwood).

McDONNELL, L. and PASCAL, A. (1979) *Organized Teachers in American Schools* (Santa Monica: Rand).

McGIVNEY, J. H. (1984) State educational governance patterns, *Educational Administration Quarterly*, 20(2): 43–63.

McLAUGHLIN, M. (1987) Learning from experience: lessons from policy implementation, *Educational Evaluation ar ! Policy Analysis*, 9(2), 171–178.

McLAUGHLIN, M. and BERMAN, P. (1978) *Federal Programs Supporting Educational Change, Vol. 8: Implementing and Sustaining Innovation* (Santa Monica: Rand).

MEIER, K., STEWART, J. and ENGLAND, R. (1989) *Race, Class and Education* (Madison: University of Wisconsin Press).

MITCHELL, D. E. (1988) Educational politics and policy: the state level, in N. Boyan (ed.) *Handbook of Research on Educational Administration* (New York: Longman), 453–466.

MITCHELL, D. E. (1990) Education politics for the new century: past issues and future directions, in D. E. Mitchell and M. E. Goertz (eds) *Education Politics for the New Century* (London: Falmer Press), Ch. 10, 153–167.

MURPHY, J. T. (1971) Title I of ESEA: the politics of implementing federal education reform, *Harvard Educational Review*, 41: 35–63.

NYSTRAND, R. O. (1976) State education policy systems, in R. Campbell and T. Mazzoni (eds) *State Policy Making for the Public Schools* (Berkeley: McCutchan), 254–264.

O'DAY, J. A. and SMITH, M. S. (1993) Systemic reform and educational opportunity, in S. Fuhrman (ed.) *Designing Coherent Education Policy* (San Francisco: Jossey-Bass), Ch. 8, 250–312.

ORDESHOOK, P. C. (1986) *Game Theory and Political Theory* (New York: Cambridge University Press).

ORFIELD, G. (1969) *The Reconstruction of Southern Education: The Schools and the 1964 Civil Rights Act* (New York: Wiley).

ORLAND, M. (1993) From the picket to the chain link fence: national goals and federal aid to the disadvantaged, Working Paper (Washington, DC: National Education Goals Panel).

PETERSON, P. E. (1974) The politics of American education, in F. N. Kerlinger and J. Carroll (eds) *Review of Research in Education*, Vol. 2 (Itasca: F. E. Peacock), 348–389.

PETERSON, P. E. (1981) *City Limits* (Chicago: University of Chicago Press).

PETERSON, P. E. (1985) *The Politics of School Reform 1870–1940* (Chicago: University of Chicago Press).

PETERSON, P. E., RABE, B. and WONG, K. (1986) *When Federalism Works* (Washington, DC: Brookings Institution).

PRESSMAN, J. and WILDAVSKY, A. (1973) *Implementation* (Berkeley: University of California Press).

RAVITCH, D. (1974) *The Great School Wars* (New York: Basic Books).

REED, R. (1992) School decentralization and empowerment, in J. Cibulka, R. Reed and K. Wong (eds) *The Politics of Urban Education in the United States* (London: Falmer Press), 149–165.

RIKER, W. (1962) *The Theory of Political Coalitions* (New Haven: Yale University Press).

ROGERS, D. (1968) *110 Livingston Street* (New York: Random House).

ROSSELL, C. (1990) *The Carrot or the Stick for School Desegregation Policy* (Philadelphia: Temple University Press).

SCRIBNER, J. D. and ENGLERT, R. (1977) The politics of education: an introduction, in J. D. Scribner (ed.) *The Politics of Education*, the Seventy-sixth Yearbook of the National Society for the Study of Education (Chicago: University of Chicago Press), 1–29.

SHEPSLE, K. A. (1989) Studying institutions: some lessons from the rational choice approach, *Journal of Theoretical Politics*, 1: 131–147.

SMITH, M. S. and O'DAY, J. A. (1991) Systemic school reform, in S. Fuhrman and B. Malen (eds) *The Politics of Curriculum and Testing* (Bristol, PA: Falmer Press), Ch. 13, 233–267.

SPRING, J. (1988) *Conflict of Interests* (New York: Longman).

STONE, C. (1976) *Economic Growth and Neighborhood Discontent* (Chapel Hill: University of North Carolina Press).

WEERES, J. and COOPER, B. (1992) Public choice perspectives on urban schools, in J. Cibulka, R. Reed and K. Wong (eds) *The Politics of Urban Education in the United States* (London: Falmer Press), 57–69.

WIRT, F. and KIRST, M. (1982) *Schools in Conflict* (Berkeley: McCutchan).

WONG, K. (1989) City implementation of federal antipoverty programs: proposing a framework, *Urban Resources*, Spring: 27–31.

WONG, K. (1990) *City Choices* (Albany: State University of New York Press).

WONG, K. (1992) The politics of urban education as a field of study: an interpretive analysis, in J. Cibulka, R. Reed and K. Wong (eds) *The Politics of Urban Education in the United States* (London: Falmer Press), 3–26.

WONG, K. (forthcoming) Can the big-city school system be governed?, in P. Cookson and B. Schneider (eds) *Creating School Policy: Trends, Dilemmas, and Prospects* (New York: Garland).

WRIGHT, D. S. (1982) *Understanding Intergovernmental Relations*, 2nd edn (Monterey, CA: Brooks/Cole).

ZEIGLER, H. and JENNINGS, K. (1974) *Governing American Schools* (North Scituat, MA: Duxbury).

ZEIGLER, H., JENNINGS, K. and PEAK, G. W. (1974) The decision-making culture of American public education, in C. Cotter (ed.) *Political Science Annual* (Indianapolis: Bobbs-Merrill), 177–226.

PART 2
The political arenas of education

3. *The crucible of democracy: the local arena*

Laurence Iannaccone and Frank W. Lutz

Introduction

For much of our history, the governance of public education has been largely left to local school districts. Over the years, the state and federal governments have appropriated increasingly greater portions of educational policy making from local school boards. This chapter focuses on the politics and governance of local school districts. It summarizes three major theoretical perspectives and relevant research in this area.

Throughout its history, research in the politics of education has been influenced more by the political realities of its day by its research knowledge and scholarly theories (Iannaccone and Cistone 1974). A number of political theories have been applied to local education governance. Among the best known and most used theories are: decision-output (input–output) systems theory, competition/participation theory, and dissatisfaction theory.

Each of these theories with its related sets of concepts paints a somewhat different picture of local school governance. These variations reflect dissimilarities in referents and in temporal assumptions. Each theory employs different methods. Each contains contrasting definitions of democracy which lead to different criteria of evaluation. Theories can provide different yet correct views of the same phenomenon. No theory is complete. Danzberger *et al.* (1992) assert that there is no one best way to improve school boards. However, theory can explain processes, make differences apparent, and suggest conceptual solutions. We will now examine and critique three of the most popular theories of local school governance. We begin with decision-output theory.

Theoretical model: decision-output theory

This theory, promulgated by Wirt and Kirst (1992), is an adaptation of Easton's (1965) framework for political analysis. Easton defined politics as the authoritative allocation of value preferences. Decision-output theory examines the relationship between inputs (resources and demands) to the political process and outputs (policy and programs) of that process as well as the subsequent outcomes of such outputs. Wirt and Kirst find that school policies and programs are seldom commensurate with citizen demands. The underpinning theory, essentially economic in nature, builds upon its central assumption of a finite pool of resources and an infinite or very near infinite craving of public interests and demands upon those resources.

Wirt and Kirst see the referendum as a critical component of the political process but find it falls short of its democratic promise in school governance. 'The promise of referendum control by citizens [in school governance] . . . has not been matched by

0268-0939/94 $10·00 © 1994 Taylor & Francis Ltd.

reality . . . Yet the promise is not completely hollow . . . these devices can be resorted to if their [policy makers'] actions are too offensive' (1992: 222–223). Wirt and Kirst conclude that the referendum is more significant for education than for other areas of public policy.

While conceding that citizens are occasionally able to compel their will at referenda, decision-output theory still views local education governance as undemocratic. While the picture so presented is accurate, it is also too limited, especially in its temporal features. As such, it fails to capture important aspects of local district political dynamics. What Wirt and Kirst view as a failure of voters to use the voting power vailable to change schools may be more accurately viewed as evidence that voter choice in these districts is clearly opposed to change. As Cusick (1992) states, 'it makes little sense to criticize a system that so well reflects society . . . coalitions of interested and appreciative citizens, students, teachers, and administrators keep the school stable. They also make the school hard to change [but] . . . not unresponsive' (pp. 227–229).

The bulk of the medium and smaller school districts generally find a way to meet the needs and satisfy the values of the citizens they serve. However, implicit in this positive judgment is the conclusion that the largest school districts where such mechanisms usually do not even exist cannot be democratically governed. In large districts such as Los Angeles, the existence of such mechanisms is meaningless because the mobilization of resources needed to use them is beyond the reach of most citizens.

Theoretical model: competition/participation theory

Zeigler *et al.* (1974) depend upon continuous competition and participation in the political arena as the major measures of democracy. They found that competition for school board membership was virtually nonexistent, describing the 'pathways to board membership . . . [as] apolitical, circumscribed, and insulated . . .' (p. 36). Further, 'the early educational reformers, those who wanted to keep politics and education separate, succeeded all too well' (p. 52) and 'the prescribed norms of democratic leadership selection run in the contrary direction of those seen in school governance' (p. 71).

Using a large amount of data on voter turnout in school board elections and the competition for school board seats, the researchers found that incumbents often run for the school board unchallenged and the turnout in school elections is low, often below 15% of those eligible to vote. Their findings of low participation and absence of competition led understandably to their conclusion that American public school governance is rather undemocratic. These researchers also found that 'the less complex the district, and the higher the mass support, the more likelihood there was of finding a school board responsive to individual preference . . . To solve the racial and social ills of urban education, it is proposed that control be radically decentralized in the central cities' (pp. 92–93). This solution was offered at least 15 years before the recent efforts at site-based management.

This portrayal of undemocratic governance is based on a rather narrow slice of the political process. It isolates one aspect of a dynamic political process, focusing only on elections. More importantly, it aggregates the data in a manner that hides the much less frequent but very significant elections in which competition is high. While accurate, the picture it draws is incomplete. Competition in school board member elections is customarily low and voter turnout poor. As a result, most voters appear in effect to disenfranchise themselves by choosing not to vote much of the time. Competition/participation theory ignores the possibility that many, perhaps most, voters are tolerably satisfied with election results most of the time.

Perhaps the most significant difference among the theories is the difference in the time frames used to collect data and describe the political process. Competition/participation theory and decision-output theory use synchronic time frames. They take rather small snap-shots of political activity at a selected moment (i.e., voting day) and occasionally paste these together to make a point (i.e., on the average not many potential voters vote in school board elections). On the other hand, dissatisfaction theory takes a longer view of the political process. This theory uses a diachronic time frame. We turn now to a detailed examinations of this theory of local school governance.

Theoretical model: dissatisfaction theory

> From the viewpoint of dissatisfaction theory, continuous competition and decision output theories appear to present us with a Hobson's choice. Their respective central concepts . . . doom the local school district to disappearance as a democratic governmental unit, the one by calling for . . . nearly universal participation, the other by subjecting us to the old tyranny, for administrator representation is despotism writ small. (Lutz and Iannaccone 1978: 129)

By the end of the 1960s, two unrelated research programs, one in the East and the other initially in the Midwest, thence moving to the West Coast, opened a significant, rather similar, window to the politics of education by examining the dynamics of politics change over time in local school districts. Unfortunately, neither of these efforts became aware of the other until much later.

McCarty and Ramsey at Cornell University produced a Community School Board Typology based on their study of local school district governance (McCarty and Ramsey 1871). This research produced a picture of systematic consistency in the relations of the community social system, school board operational style, and superintendent pattern of behavior in middle-size cities and smaller school districts. These districts reflect the majority of school districts in the United States. McCarty and Ramsey found a relationship between the type of community power structure, the school board type of internal relations, and the superintendent's role behavior.

The discovery that the community's social system is rather closely reflected in the school board's processes is not surprising. It reflects what a representative system is supposed to do. As Cusick (1992) reasoned, it makes little sense to criticize a system that so well reflects society. McCarty and Ramsey were aware of the fact that their research had not adequately probed into the dynamics of change over time in these relations. They pointed out that community power structures are impossible to classify into rigid categories and are constantly changing; what may be accurate enough at any given time may be quite inaccurate in the future (McCarty and Ramsey 1971).

McCarty and Ramsey's caution about the meaning of their cross-sectional data set and their future discussions of the dynamics interplay of the forces at work in these relationships are a tribute to the quality of their scholarship. They note that their data represent only a slice in time of a highly dynamic and changing structure. As they reported in the study:

> Originally the intention was to select a probability sample within the states. However, as we proceeded with our field research we discovered that we were not identifying a sufficient number of factional communities. In reflecting on this, we decided that factional communities often restore themselves to one of the other three types . . . Therefore, unless the researcher is in the community at the time a factional dispute is occurring, he runs the risk of overlooking the factional structure. (McCarty and Ramsey 1971: 243)

Several significant aspects of their research complement or parallel work that Lutz and Iannaccone had begun at Washington University in St Louis and that continued at the Claremont Graduate School in 1964–66 (Iannaccone 1967). In effect, though each was independent and unaware of the other, the findings of McCarty and Ramsey could easily have been included in the theoretical argument made at Claremont in 1964–66. That argument later came to be known as dissatisfaction theory (Lutz and Iannaccone 1978).

The development of dissatisfaction theory began with the Lutz dissertation case study at Washington University at St Louis in 1962. The research program began with an ethnography of the political process in a school district that included 25 years of historical data, an 18 month in-depth participant-observer-based diary, and a three-year follow-up case grounded on careful notes of every school board meeting during that three-year period.

The resulting theory argued the existence of a developmental pattern over time as a characteristic dynamic of school district governance. It posited a pattern of alternating periods of close adjustments between the school district government and the district's social composition, similar to the McCarty and Ramsey findings, gradually diverging to very distant one producing politicization and abrupt policy changes. These changes were seen as briefer periods of sharp conflict. Obviously, this demanded a longitudinal design, one that could serve as a functional equivalent in critical aspects to the long living in a community that McCarty and Ramsey had suggested was needed. Lutz and Iannaccone chose to look at the continuing process of political behavior in diachronic time as opposed to stop-action time frames of synchronic time. Thus, at any particular time, public school governance may lack participation and appear noncompetitive.

Lutz and Iannaccone found that when the policies of the board and superintendent policy-making group become too different from the community, incumbent board members will be defeated and superintendents replaced by outsiders with a new mandate to provide the services demanded by the district's voters. They argued that such a difference was most likely to occur as a result of changes in the social composition of the community without appropriate timely adjustments of the school district's educational services.

The early research on dissatisfaction theory, from 1962 to 1972, did not attempt to probe empirically the full range of the theoretical argument. The first studies went to the core political dynamic of the argument: the defeat of incumbent board members and the involuntary turnover of superintendents followed by outside rather than inside successors with the antecedent condition of changes in the social composition of the school district. Subsequent research during the next two decades has replicated the early findings and supported more of the theory's arguments.

What has emerged from this research is a view of politics as the process by which persistent and pervasive social conflicts in a polity are translated into its policy. When the policy-making core group of superintendent and board members becomes unresponsive to the demands and values held by active district voters, the officials will be replaced by the voters. This is the political dynamic of most American polities.

Dissatisfaction theory notes the lack of competition and low voter turnout in the majority of school elections. It confirms the district's unresponsiveness to citizen demands which is the result of the unrepresentativeness of 'élite' school boards. These boards view themselves as separate from and trustees for the people (see Bailey 1965). However, dissatisfaction theory also documents the periodic, dynamic processes that create more competition, increase voter turnout, create more participation at board meetings, and lead to incumbent defeat, superintendent turnover, and policy change.

Dissatisfaction theory uses a diachronic time frame and holds up the governance of local schools as a model of grass-roots democracy. It is recognized that there are times when participation and competition in local school governance are low. There are times when the policy process appears to be undemocratic. At such times, one can despair about the health of democracy in local school governance. However, the democratic process is there. To paraphrase Lutz and Merz (1992), the people can get what they want and, therefore, probably get what they deserve.

The evidence of over 30 years of research is clear. When voters in typical school districts become dissatisfied enough, they act. The sleeping nonvoting giant is awakened. Democracy is enlivened, incumbents are thrown out, superintendents are fired, and value-laden policy initiatives are redirected to become more commensurate with the demands of the people. This cycle repeats itself whenever the policies of the school board and superintendent fail to remain aligned with the wishes of voters. When serious disagreements arise, voters become motivated enough to act to realign school governance with local citizen demands. When functioning properly, this system is the embodiment of democracy. However, there is some suspicion that this system does not work as well in large urban districts. Since most reform initiatives are directed toward these districts, we need to examine why these reforms so often fail to take root.

Democratic participation and change in large urban districts

The local school district has functioned as the pivotal public school governance unit throughout most of our history. Americans have long held the notion that public education is a major means, for many perhaps the primary institution, of upward social mobility. From Jefferson through Counts (1932) to Adler (1982), American intellectual leaders have seen and continue to see schools as the cardinal organization of civic education and socialization. Voters still seem to have a deeper commitment to their local public schools than to any other governmental unit.

In reviewing the literature related to this theme, Popkewitz (1991) found that:

> ... the school has long been viewed as an essential element in the millennia vision of the United States ... the dreams of democracy, material abundance, and spiritual contentment ... depend on the success and progress of schooling. (p. 148)

Public schools continue to enjoy the support and carry the hopes of Americans as the institution looked upon to remedy the country's social problems. The myth of local control is embedded in the culture and values of the people, who tend to believe that control of 'their' public schools is and should be a matter of local control. Supporting that historical tradition, Gross et al. (1962) wrote:

> America has always cherished a belief that face-to-face democracy, the democracy of the small town, the democracy of the town meeting, is a cornerstone of the good life ... Nowhere has this social philosophy revealed itself more clearly than in our faith in the local public school and the local school district ... the general guiding principle must be, 'Keep the schools and the government of the schools close to the people so that the citizens generally ... may know what their schools are doing, and may have an effective voice in the school program ...' (pp. 78–79)

The point of departure for understanding the local district's present governance model – its essential bent, its character bias, its fundamental inclination – is the municipal reform movement. Between 1890 and 1920, the American educational policy-making system was fundamentally restructured. The manifest function of the reform movement of the early 1900s was to remove corruption from city government. One of its latent functions was to effectively separate school policy making from the poorer neighborhoods and school government from general-purpose government.

By the 1920s, the reform movement had (1) changed the system of representation in the urban school district to either appointive boards or relatively small boards elected at large; (2) changed the social class and ethnic control of urban schools; (3) established a new anti-Jacksonian pro-bureaucratic ideology and political myth about the governance of education; and (4) set into motion the forces of continuous, unlimited bureaucratic growth and the centralization of educational policy making and governance.

The reform of local school boards divorced them from general-purpose government. It reduced the numbers of board members and moved from ward-elected boards to appointed, 'blue ribbon' nominated, and/or at-large elected boards (Kirst 1994). According to Lutz (1984), this and other trends are part of the reform politics that has been the general trend for half a century, in effect removing public education from the people. Cronin (1973) stated that, by 1920, the reform had been accomplished and 'the number and kinds of school boards had diminished . . . [but] the large cities continued to search for a system that would insulate the schools even further from . . . the sordid side-effects of city politics' (p. 116).

Machine politics and school governance

The reform movement's agenda was designed to weaken the traditional big-city machine. The removal of the machine reduced some forms of corruption from city government and separated the governance of public education from general-purpose government. One consequent effect was to disconnect the general services previously administered by general-purpose government and the city political machine, e.g., legal, social, and health services desperately needed by the children of the minority poor. In their classic analysis of city government, Banfield and Wilson (1963) contend that:

> True the city machine was corrupt and sometimes vengeful. But it also provided jobs, food, fuel, rent money and, most of all, friendship to the poor and under-educated minorities of the cities. And these minorities willingly exchanged '. . . votes for [that] friendship . . .' (p. 117)

Banfield and Wilson conclude that:

> Even though in the abstract one may prefer a government that gets its influence from reasonable discussion of the common good, [or] . . . government by middle-class to government by lower-class standards, [or] . . . rule of professional administrators to that of politicians, he may nevertheless favor the machine in some particular concrete situation. (p. 127)

The authors note further that it was always the poor, under-educated minorities who favored the machine.

The reform movement tended to disenfrancise people in poorer neighborhoods. Its 'élite trusteeship' composed of at-large elected or appointed school board members was almost exclusively drawn from the upper-middle and upper classes. Such blue-ribbon boards appointed a university-trained superintendent who was hired to administer policy set by the board. Callahan (1975) documented the controversy between school boards and superintendents that led to the 'agreement' intended to leave the superintendent free from board interference when administering policy enacted by the board. That agreement remains enshrined in the culture of school boards (Lutz 1975). The effective disenfranchisement of poor, often immigrant and black urban neighborhoods was the deliberated result of social Darwinian thought. The reform effectively and deliberately disenfranchised these neighborhoods because the residents were viewed as less worthy citizens (Cubberly 1916).

Further decreased representativeness

Today's school districts are not what they were in 1920; they are an overblown extension of what the reform movement produced. Each move toward reforming school boards has taken them a further step away from the people whose children attend public schools. For more than a century, there has been pressure at the state level to decrease the number of school district via consolidation. The characteristic policy rhetoric that fueled the politics of local district consolidation during most of this century paid tribute to the dominant efficiency value of the municipal reform.

Regardless of the criteria used, e.g., teacher/pupil ratio, courses offered, availability of specializations, or cost of materials, the policy premise was that increasing the size of the district would produce more efficient schools and save money. In state after state, the number of school districts declined, their size increased, and the span between their school boards and their citizens became greater. It would, however, take a naive politics of education researcher not to see that larger districts also meant job enhancement and increased salaries for administrators and support staff. Consolidation reflected two forces: (1) organized professional demands and (2) demographic mobility toward urbanization of the whole society.

In 1932, there were 127,520 local school boards in the United States. Each of these had a local school board to establish school policy and represent the local citizenry. The political drive to consolidate school districts had begun at least 50 years earlier. The 1932 figures reflect a fraction of the number of boards representing citizens at the turn of the century. In 1939–40, there were 117,108 districts but only 15,173 in 1991–92 (US Department of Education 1993a). In 1940, there was a total population of less than 135,000,000. By 1980, the population had risen to 250,000,000. Fewer and fewer people were making the political decisions and allocating values for more and more of the populace.

These data become more significant when we note that the majority of students in the USA attend urban/city schools while a majority of school boards are located in small town/rural districts. By 1960, there were less than one-third the number of boards that existed in 1932. And by 1992, there were 15,360 or only about 12% of those which were governing in 1932 (US Bureau of the Census 1993). During this same period, public school enrollment rose from 31.4 million in 1930 to 45.3 million in 1990 (US Department of Education 1993b). This represents an increase of 44% in enrollment, using enrollment as an approximation of the public school constituency. In purely statistical terms, a 44% increase in the number of pupils with an 87% drop in the school boards elected to represent that constituency spells a severe decline in the ratio of representatives to those represented.

Since there are fewer school districts, board members represent larger constituencies. The calculus of representation shows that it is easier for constituents to get a response from their state legislators or members of Congress than from urban board members. As a result, governance by local school boards has moved further from its democratic ideal. Finally, the élite social system norms that characterize educational governance arrangements overwhelmingly reinforce the numerical disparities, making representation an absolute sham in many urban districts.

School board constituencies

A second hindrance on the ability of elected officials to represent a constituency is the degree of cultural homogeneity of the constituency they represent. It is easier to represent a constituency whose values and demands are similar than to represent one with diverse, often competing, values and demands. In 1960, approximately 13% of the total school population was characterized by race as black or 'other' (US Department of Education 1993b). By 1990, approximately 35% of the school-age population consisted of minority students (Danzberger *et al.* 1992).

These data indicate the difficulties of political representation owing to the increasing diversity of the constituencies within the largest urban districts. The greater the diversity within a single local governmental unit, the more difficult the task of representation. In sum, the number of constituents to be represented has vastly increased and the social composition of the districts has become more complex and diverse, while the number of elected officials available to represent them has decreased.

The composition of school boards

A third constraint on representation is the gap between the social-class values of most board members and those of the constituents in urban districts. In recent years, this gap has widened. Since social and political values are positively related to social class, the social-class composition of school boards is a matter of political significance. By the middle of the 1920s, municipal reform had radically changed the social composition of school boards. The middle and upper middle class had replaced the poor, ethnic neighborhood representatives on urban school boards.

The outcome of that reform persists today. The American School Board Association (1993) stated, 'the most notable characteristic of school boards is their stability or lack of change.' Between 1982 and 1992, there was almost no change in the SES demographics of the 'typical' school board. The social class characteristics of today's school board members are similar to those emerging from the reform of the mid-1920s.

However, the economic demographics of the constituency has changed dramatically. The total number of families living below the poverty level increased from 7.2 million in 1960 to 13.8 million in 1991, an increase of nearly 50% (US Department of Education 1993b). There is no evidence that current social policies will soon alter this trend. As Usdan (1994) succinctly put it:

> A major factor, of course, is the changing demographics of children and families. Indeed, the growing evidence of children's poverty is among the most salient issues facing not only the schools . . . but the society at large . . . 5.6 million children under age 6 were living in poverty in 1991 – a 33% increase . . . since 1979. (p. 375)

Notwithstanding a decade of tremendous change in the composition of the public school constituency, local school boards are still predominantly white and male (females comprise about 40% of these boards, up about 10% from their 1982 percentage). Board members are between 41 and 50 years old, earn between $40,000–$60,000 (slightly higher than 10 years before), were elected and had held office for six years (that is, they have been re-elected once on the average). Only a few big-city boards more closely mirror the populations they are supposed to represent. Needless to say, these demographics characteristics are dramatically different than those of the constituencies in urban districts.

In sum, the present system of school district representation in the largest cities is defective in at least three ways:

1. The social-class values and related political rhetoric of officials and their clients cannot mesh.
2. The ethnic and cultural fit between the governors and those governed is virtually nonexistent.
3. The calculus of representation, expressed as the ratio of representatives to voters, is an obvious sham.

The professional machine

The municipal reform movement succeeded beyond its fondest hopes in the transference of control of local districts from the neighborhood machine to the organized professional employee machines in national, state, and big city governments. The preferred politics of these machines has been described as displaying:

> ... tendencies toward: (1) the elimination of even a loyal opposition, (2) a reward pattern for maintaining the status quo, (3) the absence of adequate self-criticism, and (4) the establishment of an internal educational power elite. Nowhere at the state or local levels is this as obvious as in the case of the urban school districts, nowhere has it been as stultifying for schools as in the cities ... (Iannaccone 1967: 11)

McCarty and Ramsey's concern for school boards resulted from their assessment of these political machines:

> The greatest blow to board independence has been dealt by the phenomenal growth of militant teacher unions ... Is it possible that teacher power has rendered obsolete the old forms of control ...? (1971: 212)

Rogers similarly warned of the takeover of urban schools by the 'professional machine':

> The independence from party politics [provided by the education reform of the 1900s] should have led to more professionalism, it has not. Under the guise of professionalism a number of protective practices that are distinctly not professional have begun ... and a new form of 'educational politics' has begun. (1968: 212–213)

About the same time, the New York 'Danforth'[1] research team took note of the fact that one of the district's employee organizations, the custodian's union, was able to close New York City schools and channel hundreds of millions of school tax dollars into the pockets of its constituency (Danforth 1970). Thus, using processes reminiscent of the old political machine, the modern professional machine has filled the void left by its predecessor and rewarded its own members from the public coffers at the expense of children in urban school districts.

Governance reform or political non-event?

In the 1960s, an attempt was made to give the community a greater voice in local school district governance. The most prominent of these efforts occurred in New York City. In 1960, the New York state legislature 'reformed' the New York City school board by replacing the entire central board. The new board re-established 30 local boards. In 1967, three 'demonstration districts', funded by the Ford Foundation, were added. In 1968, membership on the central board was increased from nine to 13 members in an effort to increase racial, ethnic, and social class representation on the board. By 1969, as a result of the state reform legislation, which replaced the community control bill, the central board was reduced to five paid members and 32 local districts. None of the city's secondary schools was placed under the influence of those local boards. During the 1960s, New York

City may have been the most often reformed school district in the history of public education.

The results of this reform, however, were less than satisfactory. The Danforth report (1970) called the changes tinkering with the structure rather than reforming it. The report read, in part:

> ... the [central] board is ... incapable of responding to the society's demands in a reasonable time. It is ... a pathological bureaucracy. Above all, it follows a sacred style of governance in a secular city, relying largely on the politics of expertise ... Its well intentioned members, steeped in the traditions of the reform board movement of the early twentieth century ... have found themselves unable to respond adequately to the demands of ... the city. (pp. IV–102)

Danzberger *et al.* (1992) apparently view the New York City reform of the 1960s as an operational failure. They state that:

> ... the break up of New York City into 32 community school districts governing grades K–8 in the late 1960s did not result in improved student achievement ... The change to smaller community school districts ... has not increased citizen [nor] ... parental involvement ... (p. 83)

In sum, the New York reform of the 1960s was an attempted revolution that failed. The community control movement had the goal of transferring control of the city schools to local neighborhoods. However, its proposed legislation was not passed. Ultimately, it was defeated by the traditional power holders, led by the city's professional machines (Iannaccone 1970).

The Chicago reform of the 1990s

The decade of the 1990s has witnessed another attempt to return local control and decision making to neighborhood communities. Like New York City, the Chicago reform also seeks to break up a large urban district. However, unlike New York's community control effort, the Chicago reform seeks to restructure the city's school governance structure to the site level.

The policy premise of the present reform effort in Chicago is to shift control from traditional power holders to school site boards composed of the principal, a student, and elected citizens and teachers. This effort is more a response to political pressure and loss of public confidence than the product of educational theory; it is driven more by the demand for accountability than by pedagogical planning. Referring to this restructuring effort, Rollow and Bryk (1993) state:

> In response to the entrenched, dysfunctional power relations that had calcified ... it replaced the traditional bureaucratic control of schools with a complex of local school politics ... more responsive and accountable to their students and community ... (pp. 98–99)

According to Lutz and Merz (1992):

> The legislative action [in the Chicago reform] turned out the bureaucratic machine ... replacing it with a type of site-based management similar to the machine governance process ... The people and the parents now hold the majority of the vote on the local school councils. They run budgets and hire staff. It remains to be seen if they can produce better schools. (p. 148)

Restructuring the governance of schools to the site level in the largest cities may renew the urban school district's historic role as the crucible of democracy as Iden (1994) suggests:

... it seems to me that ... [this] is the way schools operate in small towns, small districts – as districts have gotten larger, they have tended to become less and less responsive to the citizenry whose children they serve ... it is time the communities reclaim the schools they finance. (pp. 126–127)

For site-based management to become a real, lasting change in school governance, the present educational élite – local boards, top school administrators, professional bureaucracies, state education agencies, and very possibly state and national governments – must relinquish much of their control of schools. Otherwise, site-based school boards will be no more than the failed citizen/teacher advisory committees used by the power élite for at least a half-century (Bergus 1993). Lutz and Merz (1992) warn that the reform trend toward site-based management may well result in nothing more than political rhetoric rather than real reform.

Reform or fad, only time will tell. But whatever its future, the Chicago reform stands as one more admission of the weakness of the present governance system, a product largely of the municipal reform movement of the early 1900s. As Hess (1993) reminds us:

The Chicago reform does call into question the unthinking connection between the liberal philosophical perspective and the strategies adopted by the liberals in the 1950s and 1960s, which were linked to the professionalized, rational governance strategies of the early decades of this century ... The Chicago reform effort shifts the boundaries of what is traditionally thought of as 'schooling'. Under this reform effort, schooling in urban centers is re-established in the context of local community. (pp. 92–93)

Summation and a suggestion

The recent efforts toward school restructuring suggest that we may finally have to come to realize that without fundamental changes in the governance of urban schools, reform of such schools is impossible. Significant improvement of these school demands nothing less far reaching than the reforms of the turn of the century that produced the present school governance system. As noted earlier, the reform movement of the early 1900s was the last significant school board reform movement in America (Iannaccone 1977, Danzberger *et al.* 1992). That reform no longer serves the people in the largest cities. 'The major focus of urban reform must be ... the political process and the target of that process must be the control and governance of local schools' (Lutz and Iannaccone 1993: 88).

After all the sophisticated sneering of the intelligentsia in a life-time career in world-class universities has been suffered through, all the popular plaints of political losers have been heard and all the cynical pleas on behalf of the poor by wealthy media, movie and sports millionaires has been appropriately discounted, it is still a fact that more citizens come closer to day-to-day government and have more opportunities to influence public decisions in the typical school district arena than in any other governing body. That is why Wirt and Kirst (1992) are correct when they say of the local school district voting base, 'The act of voting for or against – or not at all – links the individual citizen to the school in a direct and intimate way that is unparalleled for other major public policies' (pp. 222–223).

As noted above, the recent reforms mandating restructuring to the site level are attributable more to a response to political demands and a loss of public confidence than as the product of a careful plan for the education of the young. They are driven more by the demand for accountability than by pedagogical considerations (Weeres 1993). To the extent that restructuring programs like the Chicago reform effort become operational in the governance of site-level representative structures, they may improve accountability and enhance educational quality. But there is no guarantee that restructuring will, in fact, improve the quality of education. It may be one more case of reforming the reforms (Cuban 1990).

The results of some three decades of research indicate that most school districts have reasonably representative schools boards that produce policy consistent with active voter wishes most of the time. The research by McCarty and Ramsey, coupled with the development of dissatisfaction theory, conclude that voters change boards that do not reflect their wishes. However, the largest urban districts are another story. We have seen over three decades of efforts to increase federal control of schools and a series of urban school district reforms. The most recent and to date potentially far reaching of these reforms is school restructuring. The Chicago reform effort is the clearest example of this.

With all the other factors we usually list as 'causes' of the urban school disaster, we must include our own frames of reference. Our tendency to conceptualize all school districts, boards, and superintendents as the same sort of phenomena is erroneous. A conceptual system that places the school districts, boards, and superintendents of New York City and Elmira, Los Angeles and Santa Barbara, and Chicago and Champaign in the same category is obviously flawed. Our inability to distinguish the larger number and sorts of local districts that function as useful representative local governments from those that do not can easily lead us to an indiscriminate sameness in 'reforming' all of them together.

The body of research ignored by most of these reforms is the relatively successful political dynamics within the majority of American school districts. The policy reformer's neglected model is the one that works for most of the country. Why should we not try to use it in the places that need it most by dividing large urban districts into a number of smaller community school districts?

One objection to this is raised by the middle-class perception that reasonably sized urban communities are not feasible because they require a sense of community lacking in these areas. According to critics, the typical model will not work without that community spirit. Indeed, without that feeling of community, the model would probably not be applicable.

Most of the cities of the old West such as Los Angeles were built on a mosaic pattern unlike the concentric ring cities of the East. Urban dwellers of the Western cities tend to identify themselves as living in smaller sections and area enclaves within these large cities. This identification itself pays tribute to the communities within the Western megalopolis. But what about the rest of the country?

Summerfield's (1971) examination of urban school district politics in a Midwestern city indicates significant policy variation among the four neighborhoods he studied. While his conclusions about the openness of their politics may be challenged, the existence of distinguishable and different neighborhood education values within the larger urban system is clear. The failure of the melting pot to eliminate the identity of various ethnic enclaves in New York City, despite socio-economic successes long after the first generation of immigration, is well known (Glazer and Moynihan 1963). The retention of their identity suggests that, even in the largest cities, sub-areas of communities with their particular identity do exist and could become viable school districts. A related study of an Italian-American community in Boston and its destruction by federal urban renewal indicates another instance of internal communities with a sense of affinity in an older Eastern city (Gans 1962).

At present, Chicago is at the forefornt of efforts to restructure schools. Academics and middle-class policy makers have no difficulty in understanding the existence of community spirit and capability to govern of a Chicago community like Hyde Park. But they tend to fall back to a rabble hypothesis when thinking about the ability of the rest of the city to govern itself. The Addams community is one of Chicago's oldest slums. Suttles (1968) states, 'Seen from the inside, however, the Addams area is intricately organized according to its own standards . . .' (p. 3). The major lineaments of the area's internal structure are such customary anthropological distinctions as age, sex, territoriality, ethnicity, and personal identity. The inclusion of territoriality on this list is not merely a convention. As Suttles goes on to say, 'The most obvious reason for centering in on locality groups is that their members cannot simply ignore one another. People who routinely occupy the same place must either develop a moral order that includes all those present or fall into conflict' (p. 7).

Restructuring Chicago style may solve the educational governance problems of the largest school districts. Creating districts that capitalize on existing communities would come closer to the system that works in most of the country. Otherwise, the education failures of a few of the largest urban districts might not only terminate the reform of restructuring to the site level, but also close the chapter on the American local school district. Were this to happen, we might very well see the thruth of the statement, 'Reformers conceive of themselves as the obstetricians of the future while history declares them the morticians of the past' (Iannaccone 1967: 102).

Note

1.   The Danforth Foundation undertook the funding of three years of studies by five teams of researchers in Boston, Chicago, Columbus, Los Angeles, and New York under the leadership of Luvern Cunningham. The teams included a larger number of individuals than a note such as this can acknowledge. However, key roles were played by J. Cronin in Boston, P. Peterson in Chicago, R. Nystrand in Columbus, C. Briner in Los Angeles, and F. Lutz in New York. L. Iannaccone was initially the senior principal investigator of the New York study. He later became a floating member of the team concentrating on the state capital in the New York study. Iannaccone joined the Boston team with Cronin in the third year of the study and subsequently synthesized selected aspects of all five studies for a 1971 report to the President's Commission on Juvenile Delinquency.

References

ADLER, M. (1982) *The Paideia Proposal: An Educational Manifesto* (New York: Macmillan).

AMERICAN SCHOOL BOARD ASSOCIATION (1983) Faxed materials from ASBA report of American school boards.

BAILEY, F. G. (1965) Decisions by consensus in councils and committees, in M. Banton (ed.) *Political Systems and the Distribution of Power* (London: Tavistock).

BANFIELD, E. C. and WILSON, J. Q. (1963) *City politics* (New York: Vintage Press).

BERGUS, J. P. (1993) *Advisory councils in the public education system of the United States*, unpublished doctoral dissertation, East Texas State University (Commerce, TX).

CALLAHAN, R. E. (1975) The American board of education, 1789–1960, in P. J. Cistone (ed.) *Understanding Local School Boards* (Lexington, MA: Lexington Books).

COUNTS, G. S. (1932) *Dare the School Build a New Social Order?* (New York: John Day).

CRONIN, J. M. (1973) *The Control of Urban Schools: Perspectives on the Power of Educational Reformers* (New York: Free Press).

CUBAN, L. (1990) Reforming again, again and again, *Educational Researcher*, 19(1): 3–11.

CUBBERLY, E. P. (1916) *Public School Administration* (New York: Houghton Mifflin).

CUSIK, P. A. (1992) *The Educational System: its Nature and Logic* (New York: McGraw-Hill).

DANFORTH FOUNDATION (1970) An unpublished report on the *New York City school governance: 1964–1967*, funded by the Danforth Foundation, St Louis, MO.

DANZBERGER, J.P., KIRST, M.W. and USDAN, M.D. (1992) *Governing Public Schools: New Times New Requirements* (Washington, DC: Institute for Educational Leadership).

EASTON, D. (1965) *A Framework for Political Analysis* (New York: Prentice-Hall).

GANS, H.J. (1962) *The Urban Villagers* (New York: Free Press).

GLAZER, N. and MOYNIHAN, P. (1963) *Beyond the Melting Pot* (Cambridge: MIT Press).

GROSS, C., WRONSKI, S.P. and HANSON, J.W. (1962) *School and Society* (New York: D.C. Heath).

HESS, G.A. (1993) Race and the liberal perspective in Chicago school reform, in C. Marshall (ed.) *The New Politics of Race and Gender* (London: Falmer Press).

IDEN, R.M. (1994) Doctoral dissertation in progress at East Texas State University, Commerce, TX.

IANNACCONE, L. (1967) *Politics in Education* (New York: Center for Applied Research in Education).

IANNACCONE, L. (1970) Norms governing urban state politics of education, in F.W. Lutz (ed.) *Toward improved urban education* (Worthington, OH: Charles A. Jones).

IANNACCONE, L. (1977) Three views of change in educational politics, in J.D. Scribner (ed.) *The politics of education* (Chicago: University of Chicago Press).

IANNACCONE, L. and CISTONE, P.J. (1974) *The Politics of Education* (Eugene, OR: University of Oregon Press).

KIRST, M.W. (1994) A changing context means school board reform, *Phi Delta Kappan*, 75(5): 378–381.

LUTZ, F.W. (1975) Local school boards as sociocultural systems, in P.J. Cistone (ed.) *Understanding Local School Boards* (Lexington, MA: Lexington Books).

LUTZ, F.W. (1984) The people, their politics and their schools, *Issues in Education*, 2(3): 136–145.

LUTZ, F.W. and IANNACCONE, L. (1978) *Public Participation in Local School Districts: The Dissatisfaction Theory of American Democracy* (Lexington, MA: D.C. Heath).

LUTZ, F.W. and IANNACCONE, L. (1993) Policymakers and politics in urban education, in S.W. Rothstein (ed.) *Handbook for schooling in urban America* (Westport, CT: Greenwood Press).

LUTZ, F.W. and MERZ, C. (1992) *The Politics of School/Community Relations* (New York: Teachers College Press).

MCCARTY, D.J. and RAMSEY, C.E. (1971) *The School Managers: Power and Conflict in American Public Education* (Westport, CT: Greenwood Press).

POPKEWITZ, T.S. (1991) *A Political Sociology of Educational Reform* (New York: Teachers College Press).

ROGERS, D. (1968) *110 Livingston Street: Politics and Bureaucracy in the New York City School System* (Lexington, MA: Lexington Books).

ROLLOW, G.R. and BRYK, A.S. (1993) Democratic politics and school improvement: the potential of Chicago school reform, in C. Marshall (ed.) *The New Politics of Race and Gender* (London: Falmer Press), 97–106.

SUMMERFIELD, H.L. (1971) *The Neighborhood-based Politics of Urban Education* (Columbus, OH: Charles E. Merrill).

SUTTLES, G.D. (1968) *The Social Order of the Slum: Ethnicity and Territory in the Inner City* (Chicago: University of Chicago Press).

US BUREAU OF THE CENSUS (1993) *Statistical Abstracts of the US: The National Data Bank*, 113th ed. (Washington, DC: US Bureau of the Census).

US DEPARTMENT OF EDUCATION (1993a) *Digest of Education Statistics* (Washington, DC: National Center for Education Statistics).

US DEPARTMENT OF EDUCATION (1993b) *Common Core of Data Surveys* (Washington, DC: National Center for Education Statistics).

USDAN, M.W. (1994) The relationship between school boards and general purpose government, *Phi Delta Kappan*, 75(5): 374–377.

WEERES, J.G. (1993) The organizational structure of urban educational systems, in S.W. Rothstein (ed.) *Handbook of schooling in urban America* (Westport, CT: Greenwood Press), 113–130.

WIRT, F. and KIRST, M. (1992) *Schools in Conflict: The Politics of Education*, 3rd edn. (Berkeley: McCutchan).

ZEIGLER, L.H., JENNINGS, M.K. and PEAK, G.W. (1974) *Governing American Schools: Political Interaction in Local School Districts* (North Scituate, MA: Duxbury Press).

4.   *State policy-making and school reform: influences and influentials*

Tim L. Mazzoni

Purpose, perspective, and limitations

During the past decade, the American states have engaged in a massive use of policy in seeking to reform their public schools. Whether these nationwide efforts have had a momentous impact on education, they might well have had such an impact on governance. Certainly the states, more than ever, have become *de facto* as well as *de jure* policy makers for the schools. This chapter examines from a state – and a political influence – perspective the causes, processes, and consequences of the decade-long 'education excellence' movement. It does so by placing events in historical context, by drawing upon research findings, and by applying an open systems perspective, one concerned as much with contextual influences *upon* systems as with actor influence *within* systems.

Before beginning, two major limitations need to be acknowledged. The first is that state education policy systems are complex, shaped by external forces, and unlike one another in countless ways. The emphasis here will be on change over time more than on variability across space. The second limitation has to do with the scholarly literature (for general reviews see James 1991, Lehne 1983, McGivney 1984, Mitchell 1988). Despite advancements, there still is considerable truth in Burlingame's and Geske's 1979 conclusion that analysts have 'spent a good deal of time in some states, some time in a few states, and no time whatsoever in a great many states' (p. 60). As with variability across states, shortcomings in research limit prospects for generalization.

Expanded state activism

The last ten years have certainly witnessed an extraordinary eruption of state school policy. Though some states were vastly more aggressive than others in enacting detailed programs and broad-scale packages, the education excellence movement left no region untouched as it spread across the country. Mandating rigorous standards for students and teachers was the dominant theme of the 'first wave' of school reform (Firestone *et al.* 1991). In 1986, a 'second wave' commenced, with attention shifting from bureaucratic intensification through state prescription to school restructuring through decentralized authority. By the end of the decade, some analysts and advocates were proclaiming the beginnings of a 'third wave,' one which called for the systemic redesign of K–12 education (Murphy 1990).

0268–0939/94 $10·00 © 1994 Taylor & Francis Ltd.

Background forces

The 'why' of the 1980s policy eruption has been cast by some scholars in broad interpretive frames. Social historians (e.g., Cuban 1990, Tyack 1993) look upon these events as yet another cycle in the recurring cycles of education reform that are rooted deep in America's past. Comparative analysts (e.g., Ginsberg 1991, Plank and Adams 1989), on the other hand, identify international trends, with many countries described as seeking to utilize their school systems to cope with underlying social, political, and economic problems. Other scholars (e.g., Clark 1993, Kirst 1984, Guthrie and Koppich 1988) point to a confluence of forces in the United States, including: (1) America's slow – sometimes stagnant – economic growth, punctuated from 1978 to 1983 by soaring double-digit inflation followed by a severe national recession; (2) escalating global competition, with the loss by the United States of market share to other nations; (3) public unease and unhappiness about the ascent of economic rivals, notably the Japanese; (4) two decades of well-publicized reports decrying the softening in American schools of academic standards and the slide of student achievement test scores; and (5) élite and, to a lesser degree, popular dissatisfaction with the perceived productivity of America's public schools and inability – or unwillingness – of local school officials and educators to improve them.

Federal impetus for reform

In the context of these background forces, a political interpretation also has gained credence – namely, that the federal government during the Reagan Administration created much of the opportunity, stimulus, and agenda for state involvement on school reform issues. President Reagan was ideologically committed to the devolution of education policy, and he pushed to move program and funding responsibility from the federal to state and local levels. According to Clark and Astuto (1987), the consequence of Reagan's devolution quest, coupled with the failure of opponents to reassert a vigorous federal role, was to 'leave the territory open to the states and they are claiming the territory' (p. 71).

The Reagan Administration's exuberant use of what Jung and Kirst (1986) call the 'bully pulpit strategy' stimulated and shaped state activism. Symbolic politics replaced substantive policy as federal commissions and officials relied on evangelizing and exhortation, rather than on expenditures and enforcement, to inspire a school 'excellence' crusade (Boyd 1988). The first and most telling pulpit was afforded the National Commission on Excellence in Education, appointed by Secretary of Education, Terrel Bell. In 1983 the Commission published A Nation at Risk, a report crafted and released with a keen eye toward both arousing – and bounding – public debate on improving education in the United States (Wimpelberg and Ginsberg 1989).

Whatever its defects as policy analysis, A Nation at Risk was galvanizing as political manifesto. The imagery was one of a country in grave peril, its economy floundering and prey to foreign competitors, because a 'rising tide of mediocrity' had been allowed to erode the quality of its educational foundations. Within a year millions of citizens had heard though print and electronic media about the dismal condition of American schools and what was necessary to fix them. That the indictment of schooling was unbalanced (Bracey 1991) – and the connection between economic competitiveness and education's deterioration unexamined (Cuban 1992) – hardly detracted from the report's appeal.

A Nation at Risk was far from alone in its message. Most of the national commission reports published during the decade, and there were dozens of them, promulgated similar diagnoses and prescriptions. The influence of these reports was magnified by constant reiteration and reinforcement – amplified by extraordinary media coverage as big circulation newspapers, national magazines, and television networks discovered the schools (Kaplan 1992). Impact was further magnified by President Reagan's fervent – if belated – embrace of the bully pulpit strategy. And when, in Reagan's second term, William Bennett became Secretary of Education the reform thrust had a true 'pit bull in the bully pulpit,' a federal official whose zeal, rhetoric, and combativeness sparked one controversy – and media account – after another (Ravitch 1990: 48).

In the wake of such influences, state-level task forces, commissions, and committees did spring up across the country. Every state had at least one; many had several meeting at the same time. National influences also contributed to mounting élite and popular pressures on state policy makers to 'do something' about education. The invocation of crisis, a repeated theme in commission reports and pulpit pronouncements, infused urgency into the cause. Not only did these pressures – and the popularity to be gained by responding to them – create political incentives for state officials to risk the hazards of policy leadership, they were accompanied by education reform 'solutions' that could be readily adopted by lawmakers. The national commission reports, write McDonnell and Fuhrman (1986: 56), 'gave the impression that easily understood, simple solutions (albeit some expensive ones) were available . . . By providing, seemingly, straightforward policy solutions, these reports made it easier for state officials to propose and enact legislation quickly.'

State sources of activism

Interpretations that narrowly single out national influences give too little recognition to state activities that preceded *A Nation at Risk* and too much recognition to similarities – rather than to differences – in how state education policy systems sought in the 1980s to improve their public schools. To begin, the states had long been active on education issues. They did not need the federal government to cede them that terrain; they already occupied most of it (on state activism in the 1960s and 1970s see particularly Campbell and Mazzoni 1976, Fuhrman 1979, Iannaccone 1967, Mitchell 1981, Murphy 1982, Usdan *et al.* 1969, Wirt 1977). The federal withdrawal in the 1980s from a role emphasizing programs and funding enlarged state involvement in education, just as with other public policy domains (Nathan 1993). State government became the target of expectations, demands, and interests that could count on little but symbolic fulfillment from the Reagan Administration.

Four states – Mississippi in 1982 and Florida, California, and Arkansas in 1983 – had either passed sweeping reform legislation or were well on the way to enactment before the release of *A Nation at Risk* (Alexander 1986, Jenkins and Pearson 1991, Massell and Kirst 1986, Osborne 1988). And for several other bellwether states, notably Tennessee and South Carolina which adopted big legislative packages in 1984, the antecedents of reform clearly traced back to earlier state events (Achilles *et al.* 1986, Chance 1986). Thus, it seems fair to conclude, with Pipho (1986: K1), that *A Nation at Risk* 'fell in at the head of a parade that had already begun to take shape.'

The efficiency and productivity concerns of the 1980s were not new arrivals on the state policy scene. 'Accountability' for education – and educators – had been the subject of extensive legislative and regulatory action throughout the 1970s. Only school finance reform was a bigger issue across the states. Thirty-five states were reported as having adopted accountability legislation by 1975, with emphasis being on comprehensive planning and statewide assessment programs (Kirst 1990, Timpane 1978). By the end of the decade, 39 state legislatures were reported as having passed bills requiring minimum competency testing (Pipho 1979). In California and Florida, in particular, educational accountability became a recurrent policy theme right into the early 1980s (Herrington *et al.* 1992, Kirst 1990). By that time, these states had, as Guthrie and Koppich (1988: 46) note, 'pioneered many of the proposals contained in *A Nation at Risk* prior to its publication.'

The reformist South

The first wave of the education excellence movement crested in mid-decade and by 1987 had largely spent its force (Kirst 1988). During this initial surge of policy activism, some one-third of the states enacted sweeping, multiple-initiative reforms which closely paralleled many of the recommendations being publicized in the national commission reports. And other states adopted in a more restricted, incremental fashion a number of the same measures (Pipho 1986). But similar language did not necessarily translate into similar meaning or consequence, as each state's policy response was situated in its own unique political context (Fuhrman 1989, Marshall *et al.* 1989). There was, moreover, a distinct regional cast to the start and early spread of the reform movement. Nor was the movement quite so national in its policy legacy as many commentators made it appear.

With the notable exception of California, long a pacesetter in education, the leaders in instituting first-wave reforms were all Southern states (Pipho 1986). And, after two years of activity, these states remained at the top in nationwide surveys taken of reform accomplishments, as measured by *A Nation at Risk* prescriptions (Plank 1988). In explaining the South's enthusiasm for the education excellence movement, scholars emphasize the link between economics and education. Timar and Kirp (1988), for example, argue that it was 'regional competition for high technology firms and recognition that the region's economic future depended on a skilled work force and good educational system [which] spurred a dramatic school reform movement throughout the South' (p. 97). Vold and DeVitis (1991) maintain that there was a further – and mighty – spur to Southern activism: 'discomfort with the role of perennial underachiever' (p. 2). The growing mood of dissatisfaction created a rare opportunity for reformers; there might be 'no better time . . . to make Southern schools as good as – or better than – those found in other parts of the country' (Vold and DeVitis 1991: 2).

The South was also the most conservative region of the country, one likely to be particularly receptive to the ideology permeating the national commission reports and Reagan Administration advocacy. This might explain why, for example, *A Nation at Risk* became the template for policy in conservative Alabama (Rudder 1991), while it was virtually ignored by state lawmakers in liberal Minnesota (Mazzoni and Sullivan 1986). Also, the Southern states had typically evolved a more centralized approach to school governance than had other regions of the country (Wirt 1977). Despite localism always being present – and intense in some Southern states, such as Texas and Georgia – it was not as much of a political constraint in the South generally as it was, for example, in the New England and Rocky Mountain states.

The differential impact of the 1980–82 national recession contributed to the South's being more able than other regions to focus on – and fund – costly school reform initiatives. Although not spared, the Southern states were not hit nearly as hard economically as many states in the 'Frostbelt.' In such states as Massachusetts, Michigan, and Minnesota pressing economic issues dominated policy agendas into the mid-1980s (Mazzoni and Sullivan 1986, Osborne 1988).

Finally, the adoption of school reform legislation within the South appeared to fit the regional diffusion model of policy innovation, a model in which a state's policy making is assumed to be strongly influenced by the actions of its immediate neighbors (Gray 1994). For example, a South Carolina official commented: 'We got a lot of help from Mississippi, even though their reforms are much different . . . the process we used [local fora] was similar' (Chance 1986: 53). And a Florida legislator reported that 'people in the state joked about contests between Tennessee and Florida as to who would come out first with a merit pay plan' (Chance 1986: 83). In addition, the Southern states had regional forums, notably the Southern Regional Education Board, that facilitated the sharing of ideas, proposals, and strategies.

Enabling and energizing forces

The education reform movement swelled, of course, far beyond its original moorings in California and the South. By 1990, comprehensive policies had been legislated in states as diverse as Kentucky, Missouri, Illinois, New Mexico, Ohio, Iowa, and Washington (Alexander 1990, Pipho 1986, Kirst 1988). Commission reports, pulpit exhortations, media publicity, and high-profile advocates, when taken together, were certainly a contributing – probably a necessary – cause for the nationwide diffusion of innovation. Still, they were not sufficient; at least four other forces enabled and energized the process.

The most basic enabling force was the institutional capacity that had been steadily developing in state governments for two decades. Modern, responsive, and capable political institutions were generally to be found in the American states by the early 1980s (Mitchell 1981, Murphy 1982, Rosenthal 1990). Strengthened institutional capacity enabled state governments to innovate across a range of complex issues, in which K–12 education was simply one locus of intense activity. Scores of laws were enacted in the 1980s in such areas as water quality, air pollution, technology development, and energy conservation (Doyle and Hartle 1985, Nathan 1993, Van Horn 1993). State capability might not have been up to the challenge of producing 'coherent policy' for upgrading education (Fuhrman 1993); and the intensification of pluralistic politics, another long-term trend, worked against such coherence (Johnson 1993, Mazzoni 1993). But state capability was up to the challenge of enabling lawmakers to respond to – or seize upon – demands for school reform with an unprecedented volume and variety of policy initiatives.

A second enabling factor was the return by the mid-1980s of economic prosperity. The national recession of the early 1980s had been deep and prolonged. Fiscal year 1984, however, saw a rebound in state revenues as the national economy improved and state taxes – 36 states had to hike taxes during the recession – generated ample new monies (McDonnell and Fuhrman 1986). A growing economy and state fiscal surpluses permitted reformers to pump enough money into the bargaining arena to accommodate conflicting interests. The something-for-everyone compromise, a hallmark of omnibus bills that often were vehicles for school legislation, was made possible on a broad scale by a surge of revenues flowing into state coffers.

A pervasive energizing factor was the escalating competition among American states in the 1980s to attract or retain economic resources. In this competition, having good schools – or, at least, the reputation for them – was perceived as a vital asset. State after state trumpeted the virtues of its schools as each sought to gain an edge in the new global marketplace (Doyle and Hartle 1985). Economic competition contributed mightily to the diffusion of state education reforms, a dynamic that illustrates what Dye (1990) has proclaimed to be the new 'competitive federalism model' of policy innovation in the United States.

A second energizing force was judicial intervention, largely through the resurgence at the end of the decade of school finance lawsuits and court rulings. The court order had been a prime mover of policy activism for many states during the nationwide movement in the 1970s to redress disparities in school funding (Fuhrman 1979, Odden and Wohlstetter 1992). Then, for a decade, these 'equity' concerns were submerged – though not completely displaced (West Virginia, for example, had to respond to a sweeping school finance decision) – by the 'excellence' impulse of first-wave reforms. In 1989, however, unequal funding systems in Kentucky, Montana, and Texas were struck down by the courts. In the early 1990s, the litigation momentum continued to build, with school finance systems in a number of states being overturned for violating state constitutional provisions.

The intervention of the court had mixed results. While most states since 1970 had experienced school finance litigation – 42 as of late 1993 – plaintiffs in these cases had not been much more successful than defendants – 15 rulings for plaintiffs and 13 for defendants, with 13 cases pending, as of late 1993 (Harp 1993). Still, when a plaintiff's challenge was upheld, it established an agenda priority for the policy system to which state lawmakers had to respond, however grudgingly and minimally. In some states, notably in New Jersey and Texas, this process proved to be protracted and its outcome problematic. But in Kentucky, where the state supreme court declared in 1989 the entire public school system to be unconstitutional, judicial intervention combined with gubernatorial leadership to produce in 1990 the most comprehensive approach to school reform legislated in any American state (Alexander 1990).

'Windows' and 'entrepreneurs'

Contextual pressures also reconfigured the opportunity structure for individual system actors. These pressures opened up, in Kingdon's (1984) terms, "policy windows,' and there were a host of strategically placed 'policy entrepreneurs' in and out of government ready, willing, and able to seize the moment to 'hook solutions to problems' and 'proposals to political momentum' (p. 191). Predominant among these individuals were elected officials, especially a 'new breed' of state governors who were hailed by the media and depicted by scholars as pivotal actors in promoting education reforms (Chance 1986, Fuhrman 1989, Mueller and McKeown 1986, Osborne 1988, Rosenthal 1990, Timar and Kirp 1988). Chief state school officers were prominent in several states (Chance 1986, Layton 1986, Massell and Kirst 1986, Prestine 1989). In one state – Texas – an individual businessman appears to have forcefully stamped his priorities and management philosophy on school reform legislation (Lutz 1986, McNeil 1988).

Visible leaders were not, however, the only policy entrepreneurs pushing for school reform. Individual legislators, whose activity generally received far less media coverage than governors, were influential across the states; indeed, their overall impact probably exceeded that of any other single class of actor (Fuhrman 1990, Marshall et al. 1986). Along with lawmakers, there were behind-the-scenes policy entrepreneurs among officials, managers, and specialists in state education agencies. In some states, these 'bureaucrats' took advantage of the agenda prominence of school reform to put forward their preferred solutions and maneuver them into enactments (Chance 1986, Layton 1986, Holderness 1992, Mazzoni and Sullivan 1986).

Besides government insiders, there were individuals outside government infusing their policy systems with entrepreneurial energy. Some were linked to – and drew their influence from – national networks that had formed around such reform issues as curriculum standards, educational choice, and school-based management (Kaplan and Usdan 1992, Kirst 1984, Ogawa 1993). Others were more home grown in origin, such as the individual policy entrepreneurs in Minnesota who, working with and through a public interest group – the Citizens League – were central to moving that state along a restructuring agenda for school reform (Mazzoni 1993).

The politics of reform policy making

Process characteristics

State policy making in the 1980s on education excellence issues was not politics as usual, at least not in the states where a sweeping array of reform initiatives was undertaken. In such 'high-change' states 'much of the tactical plan,' Chance (1986: 29) observes, 'was intended to limit, control, surmount, circumvent, or avoid the constraints of more conventional processes.' Reform politics, usually in a short burst of extraordinary policy energy, supplemented or supplanted regular politics. Education policy making transcended traditional subsystem arenas – and their specialized and established legislator, bureaucrat, and lobbyist actors – and played out in broader, more public arenas.

In reform politics, top-level leaders – governors, legislators, and Chief State School Officers (CSSOs) – took charge. Blue-ribbon commissions were formed. Change initiatives of every sort were put forward, sometimes dozens were combined and compromised into a single omnibus package. High-cost and redistributive proposals were mostly siphoned out (Firestone et al. 1991). 'Political imperatives demand[ed],' in Chance's (1986: 29) words, 'slogans and easily described, cost-contained, and symbol-accommodating solutions.' Historic political alignments were often bypassed. New advocacy coalitions were forged, with the weighty influence of business groups adding impressive clout. 'Standards,' 'excellence,' and 'quality' provided unifying, motivating, and legitimating symbols. Persuasion, bargaining, and trade-offs – and, on occasion, arm-twisting pressure – mobilized supporters and neutralized opponents. The most formidable potential resisters, the big education groups, were brought aboard, bought off, or brushed aside (McDonnell and Pascal 1988). Policy visibility attracted and was expanded by media publicity. Popular support was assured; political credits were amassed; policy – and personal – agendas were attained. And a 'juggernaut' of education reforms rolled through many a state legislature (Chance 1986).

The scope, composition, and unity of the education reform movement varied. In some states – for example, Georgia and Missouri – the reform coalition reflected essentially an élite consensus (Fuhrman 1989, Hall 1989). In other states – for example, Florida, Illinois, and California – highly pluralistic politics had to be accommodated (Alexander 1986, Chance 1986, Massell and Kirst 1986). In a few states – notably Mississippi and South Carolina – coalitional efforts went beyond alignments among state-level actors and extended significant political interaction to grassroots participants (Jenkins and Person 1991, Timar and Kirp 1988). In some states – for example, California, Illinois, Georgia, and Missouri – agreement was reached among the major interests; all embraced the final compromise. Yet in other states – for example, Tennessee, Texas, and Arkansas – reform politics proved to be polarizing as political leaders and their business allies squared off in abrasive confrontation against the teacher unions and other education groups over such divisive issues as merit pay, career ladders, 'no pass/no play' rules, and teacher testing (Achilles *et al.* 1986, Fowler 1988, Lutz 1986, McNeil 1988).

Differences among the high-change states were relatively minor, however, compared with differences across the other two-thirds of the American states as they experienced the nationwide impulse to improve public schools. Several states in which CSSOs and education departments played a decisive role – for example, Colorado, Washington, Wisconsin, and New Mexico – took a critical, cautious, or containment approach to education reform (Chance 1986, Cibulka and Derlin 1992, Holderness 1992). In other states – for example, Arizona and Pennsylvania – governors developed and pressed for sweeping changes, but only modest departures were adopted by their legislatures (Karper and Boyd 1988, Osborne 1988, Sacken and Medina 1990). In still other states, governor-led reform coalitions came to focus largely on a particular initiative – for example, career ladders in Utah and educational choice in Minnesota – and succeeded, despite sometimes powerful opposition, in passing breakthrough legislation (Malen and Campbell 1986, Mazzoni 1988). And, finally, there were some states where the waves of reform washed over the education policy system with little discernible impact on decision-making processes other than those associated with temporary commissions, expanded conversation, and modest innovation (Chance 1986).

Mainstream politics

The reform movement of the 1980s plunged education ever more into the political mainstream (Fuhrman 1987). While there also was an escalation in the policy activity of state boards of education – and, in some states, they took a proactive role (Chance 1986, Layton 1986) – the arenas for action were typically the legislature and the governor's office. Reform proposals by the many hundreds were picked up, packaged, and promoted by elected officials across the country. Some of these officials, most notably governors, regularly campaigned on educational issues; then emphasized them in new initiatives. Their intentions and influence decisively shaped state policy making, a process in which 'politics' as played out in general governance arenas through symbol manipulation, interest representation, coalition building, and give-and-take bargaining loomed large.

That politicization had come to characterize education policy making did not represent a deviation from prior trends. Politicization of this process had become obvious in many states during the preceding decade (Campbell and Mazzoni 1976, Geske 1977–78, Lehne 1978, Murphy 1982, Rosenthal and Fuhrman 1981, Rost 1979). It was most evident on the issue of school finance reform (Brown and Elmore 1983, Fuhrman 1979). But continuity did not mean the absence of change. Education being politicized in the 1980s was not the same as it being politicized in the 1970s. Among the changes in state school politics, three stand out: (1) the activism of governors, (2) the involvement of big business, and (3) the influence of national organizations and networks.

Gubernatorial activism

The education reform movement afforded the nation's governors a unique opportunity as well as motivating incentive to overcome policy-making fragmentation and to marshal broad-based support for innovative legislation. A group of able, ambitious, and pragmatic governors – mainly but not solely in the South – took the spotlight as 'policy chiefs' (Caldwell 1985). Focused and determined, the activist governors persisted, frequently in the face of initial setbacks, in pressing for education reform legislation. More than any other state actor, they had the institutional authority, organizational resources, and media access to dramatize need, frame issues, and set agendas. Concentrating their policy arguments on the link between a state's school system and its economic competitiveness, governors made education reform their number one legislative priority, ahead even of tax, economic, and environmental concerns (Beyle 1990).

To pave the way for their education priorities, governors appointed blue-ribbon commissions. Such a commission could serve many functions. For example, as Malen and Campbell (1986) report for Utah, it could 'mute executive-partisan tensions by becoming an umbrella organization' where proposals could be formulated and consensus forged; it could help fix public attention on and inspire support for education reforms; and, when key lawmakers were commission members, it could establish effective linkages between itself and the legislative process (pp. 265–266). Malen and Campbell point to the commission they studied as having had a 'central role . . . in the policymaking process' (p. 266). Other analysts have arrived at the same judgment about governor-appointed commissions (e.g., Chance 1986, Fuhrman 1989). 'In many of the states that underwent reforms,' Rosenthal (1990: 111) concludes, 'commissions spearheaded the drive.'

To buttress the work of commissions, many activist governors engaged in high-powered 'issue campaigns' to arouse popular sentiment (Durning 1989). Conducted like an election campaign, these usually consisted of 'a campaign organization, a campaign kickoff, a series of campaign speeches, a campaign tour, and a panoply of campaign slogans, endorsements, advertisements, and materials' (Beyle (1993: 96). The strategy, even when cast as an all-out appeal, was not always successful. Governor Perpich, for example, lost in his first run at public school choice in Minnesota (Mazzoni 1989); and Governor Schaefer could not secure legislative approval for a special math and science high school in Maryland (Rosenthal 1990). Yet other governors – for example, Winter in Mississippi, Alexander in Tennessee, Clinton in Arkansas, and Riley in South Carolina – engendered widespread support with their campaigns.

Effective education governors were back-stage actors as well, drawing upon the tactics of insider politics to strike accords with other influentials – or to persuade or pressure them into cooperating. Governor Clinton, for instance, often followed through on his grassroots campaigns by applying a 'full court press' in the legislature, cutting deals and negotiating compromises (Ehrenhalt 1993: 123). Said Governor Babbitt of Arizona: 'In any given year, I have selected . . . issues and used everything at my disposal – initiative, referendum, the bully pulpit, the press, browbeating, trade-offs, threats, rewards – to get what I needed' (Osborne 1988: 140).

Gubernatorial involvement in the 1980s certainly exceeded that in past decades. Moving beyond budgetary responsibilities and fiscal concerns, governors took on the education quality issue and, in so doing, thrust state policy deep into the core of traditional schooling concerns (Kirst 1984). On reform initiatives to which they assigned top priority, governors usually had great influence, especially when they were committed, tenacious, and accommodating (Rosenthal 1990). Even so, gubernatorial activism was complemented – or countered – by legislative activism on the part of committee chairs, other education policy specialists, and, in some states, house and senate leaders (Hamm 1989). Case study data portray legislators as the 'active pilots' in the reform process in a number of states (Fuhrman 1990). Such, for example, was the case for House Majority Leader Connie Levi on the postsecondary choice option in Minnesota (Mazzoni 1993); for Senator Anne Lindeman on career ladders for teachers in Arizona (Firestone 1989); and for House Speaker Vera Katz on the comprehensive redesign of K–12 education in Oregon (Clark 1993). Furthermore, legislators tended to become involved in the whole spectrum of K–12 concerns, while governors targeted their power on a relatively few themes (Rosenthal 1990). Legislators did most of the steady work in shaping policy; governors did the high-profile policy work, exerting a more showy influence on selected issues.

Governors were not the only executive officials advancing major proposals to reform the schools. Chief state school officers were among the key actors in many states – for example, in California, Illinois, New York, South Carolina, and Wisconsin (Chance 1986, Layton 1986, Massell and Kirst 1986, Prestine 1989). In at least a few states CSSOs openly contested governors for policy leadership. This was especially true when these agency heads were elected officeholders – in 15 states as of 1992 (McCarthy *et al.* 1993). Having their own political constituencies, support groups, and regime interests, elected CSSOs had the resources and incentives to set an independent policy course. And these powerful actors in several states, notably in California and Wisconsin, publicly clashed with governors over issues of school funding and reform (Cibulka and Derlin 1992, Kirst and Yee 1994).

Another factor constraining governors was their need for legislator backing to get bills enacted. Executive–legislative conflict became so bitter by the end of the 1980s that it constituted in some states a political 'war between the branches' (Rosenthal 1990: 200). Conflict escalation was fueled by the growing assertiveness of legislatures, intensification of interest group pressures, media reporting focused on controversy, candidate-driven election campaigns, abrasive partisanship, and divided government (Rosenthal 1990). The last of these – divided government – had come to affect most American states. As Beyle (1993: 91) summarizes: 'Since the mid-1980s, about three-fifths of the states had "powersplits" [i.e. governorship and legislature not controlled by the same party]; 30 states [had such splits] following the 1991 elections.'

The constant press of other state issues – for example, taxes, jobs, welfare, health care, and crime – could marginalize education reform on a governor's policy agenda. Another cause of issue displacement was economic downturn. National recession and state revenue shortfalls marked the early 1990s just as they had the early 1980s (Raimondo 1993). In some states, like Massachusetts and Michigan, issues of school reform and funding became embroiled in legislative and partisan strife (Pipho 1993), with these issues becoming a political vehicle for more overarching power struggles. Yet despite all the pressures, problems, and politicization, education reform did not disappear from gubernatorial agendas. Governors have showed much more staying power on that issue than would have been predicted from past performances, though they remained more episodic actors than were committee chairs and other education policy specialists in state legislatures.

Big business involvement

The most dramatic political change associated with the 1980s reform movement was the emergence of corporate executives, organizations, and networks as education policy actors. Big business, prior to 1980, had not sought such policy involvement. Participation was largely confined to school finance issues, with the typical reaction being one of opposition to 'expensive' state reforms. Business interests also were active in pressing for tax limitation measures, which in states like Massachusetts crippled public school support (Timpane 1984). Whether motivated by self-interest related to workforce needs, by a sense of crisis rooted in vulnerability to global economic competition, or by a belief in corporate civic responsibility (McGuire 1990), big business entered state education policy arenas in full force in the 1980s.

State Business Roundtables, or like organizations, set up task forces, special commissions, and study committees across the country (Borman *et al.* 1993). Corporation executives also served on these bodies when they were created by political leaders. Business-sponsored studies were conducted, proposals were put forward, money was solicited, lobbying was undertaken, and public relations campaigns were mounted. The business community in many states became a core member of the coalition advocating education reform and a central contributor to that coalition's political influence.

The evidence that big business was a new and significant actor in state school policy making is compelling (Berman and Clugston 1988, Borman *et al.* 1993, Chance 1986, Fuhrman 1989, Massell and Fuhrman 1994, Mueller and McKeown 1986). Yet, just as with governors, the impact of business as an education policy actor should not be exaggerated. These interest groups were not uniformly influential on K–12 policy issues across the states, nor were they as unified or powerful in comparison with other system actors as might be supposed, given the tremendous resources and privileged access of corporate enterprise within the American polity (Lindblom 1977).

In a few states, the power of big business does appear to have decisively shaped K–12 reform legislation. In Texas, this influence largely emanated from computer executive H. Ross Perot who steered – or steamrollered, depending on the account – that state's 1984 school reform package into law (Chance 1986, Lutz 1986, McNeil 1988). In Georgia, business officials representing the state's multinational corporations were portrayed as exercising dominant influence on the governor-appointed commission which paved the way for that state's 1985 Quality Basic Education Act (Fuhrman 1989: 66–67). Still, these were not typical patterns. Even in states such as Arizona, Arkansas, Kentucky, Florida, and South Carolina, where business influence was clearly very substantial, it was hardly controlling (Alexander 1986, Collins 1991, Hatic and LaBrecque 1989, Osborne 1988, Timar and Kirp 1988). This was even more true in states with intensely pluralistic politics like California and, on a much smaller scale, Minnesota. The California Business Roundtable and the Minnesota Business Partnership were certainly among the key actors in moving their states toward K–12 reform (Berman and Clugston 1988). They operated, however, in intensely competitive systems where all manner of executive, legislative, agency, and interest group actors contended for policy influence (Kirst 1983, Mazzoni and Clugston 1987). In still other states – for example, Missouri, New York, and Wisconsin – the role of big business in the 1980s reforms seems to have been quite modest, at least until near the end of the decade (Cibulka and Derlin 1992, Farnham and Muth 1989, Hall 1989, Layton 1986).

As with any lobbying group, big business had to have the backing of governors and key lawmakers for its initiatives to have any prospect of passage. State political leaders were generally welcoming and supportive when it came to corporate involvement, but these officials were not the captives or pawns of business interests. Governors and legislators could – and often did – select, adapt, and reformulate proposed innovations to suit their own policy and political requirements (e.g., see Mazzoni and Clugston 1987). Political leaders not business leaders held the policy reins.

Nor were state business interests as representative or cohesive as 'the business community' label implies. While corporation executives acting through Business Roundtables and other organizations sought to rally business around the cause of reforming education, some business interests were not well represented. In particular, as McGuire (1990: 114) observes, 'small businesses and small business organizations have not become major players.' And more marginalized groups – for example, women and nonwhite business owners – evidently had little voice in the councils of business élites (Borman et al. 1993: 69). The élites who counted most were those representing America's big multinational corporations, the business sector most closely linked to the competitive demands of the global marketplace.

As a state policy actor, the business lobby was fragmented, 'even Balkanized' (Rosenthal 1993: 151). A state's business community consisted of a number of distinct interests, often in conflict and competition with one another (Thomas and Hrebenar 1990). In some states, a broad consensus was forged around a school reform plan. In other states, however, different business groups put forward different – and rival – proposals, with one important line of cleavage being over whether or not business should throw its political weight behind privatization, vouchers, and public funding of private schools (Weisman 1991). And in still other states, business interests divided sharply over whether school reform justified tax increases. Such was the case in Florida when its 1983 reform package was at issue in the legislature (Alexander 1986).

Along with being internally split, state business lobbies had to operate in highly pluralistic political environments. By the mid-1980s, state education policy systems had become congested with individuals and groups trying to set agendas and shape decisions. The mainline K–12 groups representing teachers, administrators, and boards had been joined over the decades by a myriad of other organized interests in education. In addition, noneducation groups other than business – for example, parent, civic, urban, labor, farm, and foundation groups – wanted to have a crack at changing schools. Crowded arenas and competitive politics constrained the influence of any particular group. 'There are so many interests and so much pressure in the statehouses that only a few demands go unopposed,' observes one scholar of state politics. 'Policymaking in many places is more pluralistic than before and any single interest is less likely to dominate' (Rosenthal 1993: 216).

Thus, for all its impressive resources – wealth, status, organization, access, and 'strategic position,' among them (Hayes 1992: 49) – big business was not generally the most influential interest group operating in state education policy systems. When it got down to head-to-head conflict in the legislature, as it occasionally did on such issues as career ladders, testing of teachers, and school choice, the counterveiling power of the teacher unions and other education interest groups significantly constrained the policy-making influence of big business as well as that of other reform proponents (Cibulka and Derlin 1992, Fiore 1990, Fowler 1988, Mazzoni 1988, Rudiak and Plank 1992).

Two major studies done in the 1980s provide attributional data as to the relative influence of big business, teacher unions, and other interest groups in state policy making. Comparative research by Marshall et al. (1986) in Arizona, California, Illinois, Pennsylvania, West Virginia, and Wisconsin involved extensive interviews with education policy actors – legislators, executive members, agency administrators, lobbyists, etc. As part of the study, respondents rated the education policy influence of 18 state actors. The overall assessments for 1982–85, when put into rank orders, portrayed 'all education interests groups combined' (ranked fourth) and 'teachers' organizations' (ranked fifth) as being in what the study authors call the Near Circle of power (pp. 351–352). These groups ranked behind only legislators and CSSOs as policy influentials and ahead, surprisingly, of governors (ranked sixth). As for business organizations, they were considered under the rubric 'noneducation groups,' a category of actor which in the composite rankings finished a distant 14th. In just one of the six states – Arizona – was there a ranking (fifth) which placed business interests in the Near Circle (p. 355).

The second study offering comparative evidence was a national survey conducted in the latter part of the 1980s by Thomas and Hrebenar (1990: 144–145), updated for each state in 1989. In this reputational study, which was of a general nature and did not specify any particular policy domain, the 'school teachers' organizations' were more frequently identified by the participating political scientists in each of the 50 states as belonging in the 'most effective' category (in 43 states) than any other interest group. Ranked second were 'general business organizations,' being placed in the most effective category in 31 states. Study authors (Thomas and Hrebenar 1991) conclude that business, 'despite its fragmentation,' was the 'most widespread and powerful interest active at both the national and state levels' in the United States (p. 74). They go on to add, however, that 'overall in the 50 states the most prevalent, active, and influential interest is education, especially . . . the state-level education association' (p. 75).

National organization and network influences

State education policy systems by the 1990s had become enveloped and interpenetrated by national organizations and connecting networks, an expansion of influences that had been evolving for decades. The professional associations, the oldest networks, had sustained impact on K–12 policy making dating back into the last century (Iannaccone 1967). And a school finance network of 'academic scribblers' was pointed to some three decades ago by researchers as having 'enormously influenced the course of educational policy throughout the Northeast – and beyond' (Bailey and others 1962: 24). In the mid-1960s the first nationwide compact for education – the Education Commission of the States – was formed; and this 'network of networks' (Kaplan and Usdan 1992: 671), though rarely a major state-level policy influence, had fostered much dialogue and many connections over the years among its political and education constituencies (Layton 1985). In the early 1970s, an amply funded and tightly organized advocacy network was created by the Ford Foundation, working with the federal government's National Institute of Education, to champion school finance reform in targeted states across the country (Kirst, Meister, and Rowley 1984). By the end of the seventies the 'equity network' was identified as an actor in 11 of the 28 states which overhauled their school funding formulas (Odden 1981). Other national organizations and networks were also pointed to during that decade as exerting agenda-setting influence on the issues of collective bargaining, minimum competency testing, and scientific creationism (Kirst et al. 1984). During the 1980s there was continued proliferation of these organizations and networks, constituting by the close of that decade a 'web of coalitions and advocacy groups . . . ubiquitous in shaping public policy' (Kaplan and Usdan 1992: 666).

In the first rank of national organizations and networks were those containing state political leaders. The Education Commission of the States, the National Conference of State Legislatures, the Council of Chief State School Officers, and other such organizations accelerated their education policy activities in the 1980s. Still, the one that clearly moved to the front in that decade – and stayed there – was the National Governors'

Association (NGA). Indeed, it was the use by state governors of the NGA that most clearly distinguished their activism on school reform issues in the 1980s from any previous period of involvement. In 1985, under the leadership of Governor Alexander of Tennessee, the NGA began for the first time to concentrate on public policy issues, with the first such issue to receive in-depth attention being education. The NGA's report *A Time for Results* (1986) gave a strong nationalizing impetus to second-wave reforms by disseminating 'restructuring' proposals across the states, by emphasizing their priority – and providing rationales – for individual governors willing to champion the cause, and by putting the staff resources as well as the prestige resources of the NGA behind them. As Massell and Fuhrman (1994: 18) observe: 'NGA can take ideas in good currency among policy specialists and professionals and grant them widespread political legitimacy. National action can be used as potential leverage for change within states.'

The NGA grasped the opportunity afforded by President Bush's 1989 Education Summit to gain greatly enhanced visibility and policy influence in setting a national education reform agenda. Whether, in Pipho's (1989: 182) words, 'the President took his cue from governors or whether key governors were able to make their agendas overlap with the White House,' agreement was reached at the Summit on the need for national goals in education. And it was an NGA-created task force that fashioned the six basic goals, which along with a set of 21 related objectives were embraced by both the President and the governors (Walker 1990: 17). The NGA also declared the commitment of each governor to review state goals and education performance 'in light of these national goals' (National Governors' Association 1990: 39). Finally, governors were well represented on the National Education Goals Panels which was to prepare annually a report on goal achievement. By 1990, then, America's 50 governors had projected their collective power on education issues into national as well as state policy arenas, an expression of influence that would have been unthinkable at the decade's outset.

A second major nationalizing force was big business. National business organizations and their executive heads played a highly visible role in stimulating, defining, and expanding the movement for school reform (McGuire 1990: 112–113). By 1990, the most powerful of these groups – the Business Roundtable (representing CEOs from the nation's largest private corporations), the National Alliance of Business, the National Association of Manufacturers, and the US Chamber of Commerce, among other business organizations – had come together to form a coalition to promote education reform nationwide. Business Roundtable executives made it clear that they intended to enlarge their influence in state policy arenas. Roundtable CEOs were to contact governors in all the states in which their businesses had close ties, with the intent of forging relationships, discussing issues, and formulating plans of action (Pipho 1990). The Business Roundtable also set forth nine criteria for identifying 'essential components of a successful education system,' and urged local business leaders to apply these as a standard in conducting 'gap' analyses in their states (Borman *et al.* 1993).

The activism of political and business networks was matched by that of the educational associations; for they, too, stepped up their policy involvement in the 1980s. The most powerful were the national teacher unions – NEA and AFT – which for years had sought to influence state-level issues. In the 1960s and 1970s, for example, these two organizations led a nationwide push to have state governments enact collective bargaining legislation. Their advocacy, according to Kirst *et al.* (1984: 14), resulted in 'the popularity and spread of collective bargaining statutes in the states.' In the 1980s, the teacher unions moved to the forefront in promoting teacher professionalism, school-based management, and other teacher-empowering components of school restructuring (Ogawa 1993).

A quite different manifestation of the influence of professional associations was that exerted by subject-matter organizations of educators. These organizations, like other education groups, were caught in a reactive mode by the first wave of school reforms and had relatively little say on policy. But starting in the mid-1980s, with the pioneering work of the National Council of Teachers of Mathematics (NCTM), this began to change. NCTM's curriculum and evaluation standards for mathematics, published in 1989, were reported as having widespread impact on state policies relating to curriculum content and teacher preparation (Massell and Fuhrman 1994). Following the lead of NCTM, other professional associations began the process of developing and disseminating curriculum and evaluation standards for their respective subject-matters. This development, which Massell and Fuhrman (1994: 14) attribute partly to 'NCTM's success' and partly to 'the rise of systemic reform ideas on the public issue-attention cycle,' created another access channel for professional as well as nationalizing influences on state education policy systems.

Besides the standards-raising thrust, organizations and individuals came together in national networks during the 1980s to champion other policy initiatives – for example, networks promoting America 2000 (the Bush Administration's school reform plan), public school choice, outcome-based education, and school-based management. The last of these was examined in a revealing study by Ogawa (1993) who found that 'a relatively small set of actors shaped and promoted school-based management in the national arena' (p. 39). The key actors were four organizations 'linked by a network . . . if only loosely so' (p. 40). The chief initiator was a private foundation, the Carnegie Corporation, and its creation, the Carnegie Forum on Education and the Economy (CFEE). Joining forces with CFEE was a political association (NGA), a teacher union (AFT), and a policy research center (CPRE). These organizations could draw upon ample, diverse, and complementary resources, and they were energized by highly motivated and politically skilled policy entrepreneurs. Network activists, among their many influence tactics, publicized reports, convened meetings, sponsored research, cultivated personal relationships, and lobbied state lawmakers.

One focal issue for contending national influences was state ballot or legislative initiatives which sought to provide public funding through 'vouchers' for parents of private school students. On this issue, a new national advocacy organization was formed in late 1993, just before California voters by a large margin rejected a hotly contested school-voucher ballot initiative. (Voucher proposals had also been voted down in 1990 in Oregon and in 1992 in Colorado.) Political linkages for the new organization – Americans for School Choice – were mirrored in its board of directors, which included two former secretaries of education, three governors, and several state legislators (Olson 1993b). The organization's stated mission was to build state-level advocacy groups, either by establishing them or by linking to existing groups. Warned its executive director: 'The California initiative will be the last time school choice will be fought on a single battlefield' (Olson 1993a: 17). Americans for School Choice planned by 1996 to have mobilized proponents around either ballot initiatives or legislative lobbying in 25 states. On the other side of the voucher issue, education interest groups and their allies had already put together blocking coalitions of formidable strength, coalitions which could draw upon the resources of state organizations, particularly the powerful teacher unions, and – if pressed – upon the financial and political resources of their national associations and networks.

Another focal issue for contending national forces was 'outcome(s)-based education' (OBE). The idea of redesigning K–12 systems around challenging outcomes and high standards for student performance had come by the close of the decade to hold great appeal for reformers, with such influential organizations as the NGA and Business Roundtable, plus many federal and state officials, calling for systemic approaches to school reform. By the early 1990s, legislative and state board initiatives to translate this idea into policy were under way in a variety of forms and under a variety of rubrics across the country. But in some states, beginning with Pennsylvania in 1992 (McQuaide and Pliska 1993), the specification of outcomes as the drive elements in proposed OBE systems became engulfed in heated public controversy.

Criticisms of outcome-based education (e.g., Schlafly 1993) ranged over a wide spectrum, from language vagueness and implementation costs to bureaucratic intrusiveness. Proposed outcomes that touched upon the ethics, character, or attitudes of students sparked the most angry outcry. Deep divisions surfaced, often in polarizing confrontations, over what values the schools should teach and who should determine these values (McQuaide and Pliska 1993). Much of the opposition seemed to be home grown, reflecting varied constituencies and concerns within a particular state. But national religious and conservative groups, linked through extensive networks, also became visible and vocal participants in fanning the fires of populist discontent. Outcome-based education became targeted for countermobilization in more than a dozen states by Religious Right and conservative 'pro-family' organizations, such as the Citizens for Excellence in Education, the Christian Coalition, and the Eagle Forum (Olson 1993c, Schlafly 1993, Simonds 1993). Leading the charge for outcomes-based systems suddenly became politically hazardous for elected officials, some of whom were reported to be 'backpedaling' in reaction to grassroots contacts and vociferous criticism (Olson 1993c). As assessed up by the director of a national OBE network, there was 'intense political pressure – organized political pressure – being placed on many, many districts in many, many states not to do this' (Spady as quoted in Olson 1993c).

To point to the growing involvement of national organizations and networks in state education policy making is not to say that they eclipsed the power of more proximate actors. Governors, legislators, bureaucrats, and interest groups exercised preponderant influence over most issues and over most stages of policy making (Marshall et al. 1986, Massell and Fuhrman 1994, Rosenthal 1990, 1993). These actors generally welcomed – and often actively sought out – ideas, information, and proposals from outside their borders: from other states, from their political and professional associations, and from broader policy networks. Moreover, the states varied on a host of dimensions that affected their permeability to external influences. A state's political culture was probably the most salient of these. This culture and the 'assumptive worlds' of its policy makers (Marshall et al. 1989) fundamentally shaped the impact that national organizations and networks could have on a particular state education policy system (Kirst et al. 1984).

Continuity and change

Looking back over a decade of school reform efforts suggests five concluding observations: (1) State policy activism, for all its remarkable sweep and intensity, did not mark an abrupt break with the past; (2) The policy eruption of the 1980s accelerated as well as reflected the pluralism, politicization, and openness of state education policy systems; (3) Governors and big business, usually as coalition partners, took on a vastly expanded initiating role in these policy systems; (4) Nationalizing influences increasingly shaped state education policy making; and (5) State education policy systems became arenas for political confrontation between contending national organizations and networks.

Some school reform issues, such as vouchers and outcome-based education, cut deep; they raised questions of values as well as of interests. They promised – or threatened – to institute basic structural changes, changes widely perceived as having profound redistributive implications. They could be couched in evocative, motivating symbols and slogans. They were, in short, the kind of issues around which broad-based networks could be mobilized and their partisans propelled into the political arena. That such mobilization was occurring nationwide was clear by the end of 1993. And it could make a decisive difference to the course of education reform generally in the United States, given the increasing openness of state education policy systems to outside influences, and the scope, power, and zeal of the contending organizations and networks.

References

ACHILLES, C. M., LANSFORD, Z. and PAYNE, W. H. (1986) Tennessee educational reform: gubernatorial advocacy, in V. D. Mueller and M. P. McKeown (eds) *The Fiscal, Legal, and Political Aspects of State Reform of Elementary and Secondary Education* (Cambridge, MA: Ballinger), 223–244.

ALEXANDER, K. (1986) Executive leadership and educational reform in Florida, in V. D. Mueller and M. P. McKeown (eds) *The Fiscal, Legal, and Political Aspects of State Reform of Elementary and Secondary Education* (Cambridge, MA: Ballinger), 145–168.

ALEXANDER, K. (1990) The courts and the governor show the way in Kentucky, *Politics of Education Bulletin*, 16(3); 1–3.

BAILEY, S. K. and others (1962) *Schoolmen and Politics* (Syracuse, NY: Syracuse University Press).

BEYLE, T. L. (1990) Governors, in V. Gray, H. Jacob, and R. B. Albritton (eds) *Politics in the American States*, 5th edn (Glenview, IL: Scott, Foresman), 201–251.

BEYLE, T. L. (1993) Being governor, in C. E. Van Horn (ed.) *The State of the States*, 2nd edn (Washington DC: CQ Press), 79–114.

BERMAN, P. and CLUGSTON, R. (1988) A tale of two states: the business community and educational reform in California and Minnesota, in M. Levine and R. Trachtman (eds) *American Business and the Public Schools* (New York: Teachers College Press), 121–149.

BORMAN, K., CASTENELL, L. and GALLAGHER, K. (1993) Business involvement in school reform: the rise of the business roundtable, in C. Marshall (ed.) *The New Politics of Race and Gender* (London: Falmer Press), 69–83.

BOYD, W. L. (1988) How to reform schools without half trying: secrets of the Reagan administration, *Educational Administration Quarterly*, 24(3), 299–309.

BRACEY, G. W. (1991) Why can't they be like we were?, *Phi Delta Kappan*, 73(2), 105–117.

BROWN, P. R. and ELMORE, R. F. (1983) Analyzing the impact of school finance reform, in N. Cambron-McCabe and A. Odden (eds) *The Changing Politics of School Finance* (Cambridge, MA: Ballinger), 107–138.

BURLINGAME, M. and GESKE, T. (1979) State politics and education: an examination of selected multiple state case studies, *Educational Administration Quarterly*, 15(2), 50–79.

CALDWELL, P. (1985, February 6) Governors: no longer simply patrons, they are policy chiefs, *Education Week* 4(20), 1, 34.

CAMPBELL, R. F. and MAZZONI, T. L. (1976) *State Policy Making for the Public Schools* (Berkeley, CA: McCutchan).

CHANCE, W. (1986) '... *The Best of Education,' Reforming America's Schools in the 1980s* (Olympia, Washington: MacArthur Foundation).

CIBULKA, J. G. and DERLIN, R. L. (1992) State leadership for education restructuring: a comparison of two state policy systems, paper presented at the annual meeting of the American Educational Research Association, San Francisco.

CLARK, D. L. and ASTUTO, T. A. (1987) Federal education policy in the United States: the conservative agenda and accomplishments, in *Educational Policy in Australia and America* (London: Falmer Press), 47–76.

CLARK, S. M. (1993) Higher education and school reform, *The Review of Higher Education*, 17(1), 1–20.

COLLINS, T. (1991) Reform and reaction: the political economy of education in Kentucky, paper presented at the annual meeting of the Rural Sociological Society, Columbus, Ohio.

CUBAN, L. (1990) Reforming again, again, and again, *Educational Researcher*, 19(1), 3–13.

CUBAN, L. (1992) The corporate myth of reforming public schools, *Phi Delta Kappan*, 74(2), 157–159.

DOYLE, D. P. and HARTLE, T. W. (1985) *Excellence in Education* (Washington, DC: American Enterprise Institute).

DURNING, D. (1989) Governors' issue campaigns: an exploration, paper presented at the annual meeting of the Southern Political Science Association, Memphis, Tennessee.

DYE, T. R. (1990) *American Federalism: Competition Among Governments* (Lexington, MA: Lexington Press).

EHRENHALT, A. (1993) What if a real governor became president?, in T. L. Beyle (ed.) *State Government* (Washington, DC: CQ Press), 121–124.

FARNHAM, J. and MUTH, R. (1989) Who decides, the politics of education in New York, paper presented at the annual meeting of the American Educational Research Association, San Francisco.

FIORE, A. M. (1990) The efforts of educational interest groups to defeat merit pay for teachers in Pennsylvania: 1983–1986, paper presented at the annual meeting of the Eastern Educational Research Association, Clearwater, Florida.

FIRESTONE, W. A. (1989) Educational policy as an ecology of games, *Educational Researcher*, 18(7), 18–24.

FIRESTONE, W. A., FUHRMAN, S. H. and KIRST, M. W. (1991) State educational reform since 1983: appraisal and future, *Educational Policy*, 5(3), 233–250.

FOWLER, F. C. (1988) The politics of school reform in Tennessee: a view from the classroom, in W. L. Boyd and C. T. Kerchner (eds) *The Politics of Excellence and Choice in Education* (London: Falmer Press), 183–198.

FUHRMAN, S. (1979) *State Education Politics: The Case of School Finance Reform*, with contributions by J. Berke, M. Kirst, and M. Usdan (Denver: Education Commission of the States).

FUHRMAN, S. H. (1987) Educational policy: a new context for governance, *Publius: The Journal of Federalism*, 17(3), 131–143.

FUHRMAN, S. H. (1989) State politics and education reform, in J. Hannaway and R. Crowson (eds) *The Politics of Reforming School Administration* (London: Falmer Press), 61–75.

FUHRMAN, S. (1990) Legislatures and education policy, paper presented for the Symposium on the Legislature in the Twenty-First Century, Williamsburg, VA., April 27–29.

FUHRMAN, S. H. (1993) The politics of coherence, in S. H. Fuhrman (ed.) *Designing Coherent Education Policy* (San Francisco: Jossey-Bass), 1–34.

GESKE, T. G. (1977–78) State educational policy-making, a changing scene?, *Administrator's Notebook*, 26(2), 1–4.

GINSBURG, M. B. (ed.) (1991) *Understanding Educational Reform in Global Context: Economy, Ideology, and the State* (New York: Garland).

GRAY, V. (1994) Competition, emulation, and policy innovation, in L. C. Dodd and C. Jillson (eds) *New Perspectives on American Politics* (Washington, DC: CQ Press), 230–248.

GUTHRIE, J. W. and KOPPICH, J. (1988) Exploring the political economy of national educational reform, in W. L. Boyd and C. T. Kerchner (eds) *The Politics of Excellence and Choice in Education* (London: Falmer Press), 37–47.

HALL, P. M. (1989) Policy as the transformation of intentions, act 1: Missouri's 1985 excellence in education act, paper presented at the annual meeting of the American Educational Research Association, San Francisco.

HAMM, K. E. (1989) The evolution of state legislative education committees: independent sources of power or parts of cozy triangles, paper presented at the annual meeting of the American Educational Research Association, San Francisco.

HARP, L. (1993, September 22) Momentum for challenges to finance systems still seem strong, *Education Week*, 13(3), 22.

HATIC, H. and LA BRECQUE, R. (1989) State education reform: crisis and consensus, *Educational Policy*, 3(3), 217–231.

HAYES, M. T. (1992) *Incrementalism and Public Policy* (New York: Longman).

HERRINGTON, C. D., JOHNSON, B. and O'FARRELL, M. (1992) A legislative history of accountability in Florida: 1971–1991, unpublished paper, Learning Systems Institute, Florida State University.

HOLDERNESS, S. T. (1992) The politics of state educational policymaking: usefulness of the Kingdon model, in F. C. Wendel (ed.) *Issues of Professional Preparation and Practice*, UCEA Monograph Series (University Park, PA: Pennsylvania State University), 17–31.

IANNACCONE, L. (1967) *Politics in Education* (New York: Center for Applied Research in Education).

JAMES, T. J. (1991) State authority and the politics of educational change, in G. Grant (ed.) *Review of Research in Education*, 17 (Washington, DC: American Educational Research Association), 169–224.

JENKINS, R. L. and PERSON, W. A. (1991) Educational reform in Mississippi: a historical perspective, in D. J. Vold and J. L. DeVitis (eds) *School Reform in the Deep South* (Tuscaloosa, AL: University of Alabama Press), 75–108.

JOHNSON, B. L. Jr. (1993) In search of coherent reform policy at the state level: a longitudinal study, paper presented at the annual meeting of the University Council for Educational Administration, Houston.

JUNG, R. and KIRST, M. (1986) Beyond mutual adaptation, into the bully pulpit: recent research on the federal role in education, *Educational Administration Quarterly*, 22(1), 80–109.

KAPLAN, G. R. (1992) *Images in Education: the Mass Media's Version of America's Schools* (Washington, DC: Institute for Educational Leadership and National School Public Relations Association).

KAPLAN, G. R. and USDAN, M. D. (1992) The changing look of education's policy networks, *Phi Delta Kappan*, 73(9), 664–672.

KARPER, J. H. and BOYD, W. L. (1988) Interest groups and the changing environment of state educational policymaking: developments in Pennsylvania, *Educational Administration Quarterly*, 24(1), 21–54.

KINGDON, J. (1984) *Agendas, Alternatives, and Public Policy* (Boston: Little, Brown).

KIRST, M. W. (1983) The California business roundtable: their strategy and impact on state education policy, paper prepared for the Committee for Economic Development, New York.

KIRST, M. W. (1984) *Who Controls Our Schools?* (New York: Freeman).

KIRST, M. W. (1988) Recent state education reform in the United States: looking backward and forward, *Educational Administration Quarterly*, 24(3), 319–328.

KIRST, M. W. (1990) *Accountability: Implications for State and Local Policymakers* (Washington, DC: Office of Educational Research and Improvement, US Department of Education).

KIRST, M. W., MEISTER, G. and ROWLEY, S. (1984) Policy issue networks: their influence on state policymaking, *Policy Studies Journal*, 13(1), 247–264.

KIRST, M. W. and YEE, G. (1994) An examination of the evolution of California state educational reform, 1983–1993, in D. Massell and S. Fuhrman, *Ten Years of State Education Reform. 1983–1993: Overview with Four Case Studies* (New Brunswick, NJ: Consortium for Policy Research in Education), 67–103.

LAYTON, D. H. (1985) ECS at 20: new vitality and new possibilities, *Phi Delta Kappan*, 67(4), 272–276.

LAYTON, D. H. (1986) The regents action plan: New York's educational reform initiative in the 1980s, *Peabody Journal of Education*, 63(4), 6–22.

LEHNE, R. (1978) *The Quest for Justice: The Politics of School Finance Reform* (New York: Longman).

LEHNE, R. (1983) Research perspectives on state legislatures and education policy, *Educational Evaluation and Policy Analysis*, 5(1), 43–54.

LINDBLOM, C. E. (1977) *Politics and Markets: The World's Political-Economic Systems* (New York: Basic Books).

LUTZ, F. W. (1986) Education politics in Texas, *Peabody Journal of Education*, 63(4), 70–89.

MALEN, B. and CAMPBELL, R. F. (1986) Public school reform in Utah: enacting career ladder legislation, in V. D. Mueller and M. P. McKeown (eds) *The Fiscal, Legal, and Political Aspects of State Reform of Elementary and Secondary Education* (Cambridge, MA: Ballinger), 245–276.

MARSHALL, C., MITCHELL, D., and WIRT, F. (1986) The context of state-level policy formation, *Educational Evaluation and Policy Analysis*, 8(4), 347–378.

MARSHALL, C., MITCHELL, D. and WIRT, F. (1989) *Culture and Education Policy in the American States* (London: Falmer Press).

MASSELL, D. and FUHRMAN, S. (1994) *Ten Years of State Education Reform. 1983–1993: Overview with Four Case Studies* (New Brunswick, NJ: Consortium for Policy Research in Education).

MASSELL, D. and KIRST, M. W. (1986) State policymaking for educational excellence: school reform in California, in V. D. Mueller and M. P. McKeown (eds) *The Fiscal, Legal, and Political Aspects of State Reform of Elementary and Secondary Education* (Cambridge, MA: Ballinger), 121–144.

MAZZONI, T. L. (1988) The politics of educational choice in Minnesota, in W. L. Boyd and C. T. Kerchner (eds) *The Politics of Excellence and Choice in Education* (London: Falmer Press), 217–230.

MAZZONI, T. L. (1989) Governors as policy leaders for education; a Minnesota comparison, *Educational Policy*, 3(1), 79–90.

MAZZONI, T. L. (1993) The changing politics of state education policy making; a 20-year Minnesota perspective, *Educational Evaluation and Policy Analysis*, 15(4), 357–379.

MAZZONI, T. L. and CLUGSTON, R. M. Jr. (1987) Big business as a policy innovator in state school reform: a Minnesota case study, *Educational Evaluation and Policy Analysis*, 9(4), 312–324.

MAZZONI, T. and SULLIVAN, B. (1986) State government and educational reform in Minnesota, in V. D. Mueller and M. P. McKeown (eds) *The Fiscal, Legal, and Political Aspects of State Reform of Elementary and Secondary Education* (Cambridge, MA: Ballinger), 169–202.

MCCARTHY, M., LANGDON, C. and OLSON, J. (1993) *State Education Governance Structures* (Denver: Education Commission of the States).

McDONNELL, L. and FUHRMAN, S. (1986) The political context of reform, in V. D. Mueller and M. P. McKeown (eds) *The Fiscal, Legal, and Political Aspects of State Reform of Elementary and Secondary Education* (Cambridge, MA: Ballinger), 43–64.

McDONNELL, L. M. and PASCAL, A. (1988) *Teacher Unions and Educational Reform* (New Brunswick, NJ: Center for Policy Research in Education).

McGIVNEY, J. (1984) State educational governance patterns, *Educational Administration Quarterly*, 20(2), 43–63.

McGUIRE, K. (1990) Business involvement in the 1990s, in D. E. Mitchell and M. E. Goertz (eds) *Education Politics for the New Century* (London: Falmer Press), 107–117.

McNEIL, L. M. (1988) The politics of Texas school reform, in W. L. Boyd and C. T. Kerchner (eds) *The Politics of Excellence and Choice in Education* (London: Falmer Press), 199–216.

McQUAIDE, J. and PLISKA, A. (1993) The challenge to Pennsylvania's education reform, *Educational Leadership*, 51(4), 16–21.

MITCHELL, D. E. (1981) *Shaping Legislative Decisions: Education Policy and the Social Sciences* (Lexington, MA: Heath).

MITCHELL, D. E. (1988) Educational politics and policy: the state level, in N. J. Boyan (ed.) *Handbook of Research on Educational Administration* (New York: Longman), 453–466.

MUELLER, V. D. and McKEOWN, M. P. (eds) (1986) *The Fiscal, Legal, and Political Aspects of State Reform of Elementary and Secondary Education* (Cambridge, MA: Ballinger).

MURPHY, J. T. (1982) Progress and problems: the paradox of state reform, in A. Lieberman and M. W. McLaughlin (eds) *Policy Making in Education* (Chicago: National Society for the Study of Education), 195–214.

MURPHY, J. (ed.) (1990) *The Educational Reform Movement of the 1980s* (Berkeley, CA: McCutchan).

NATHAN, R. P. (1993) The role of the states in American federalism, in C. E. Van Horn (ed.) *The State of the States*, 2nd edn (Washington, DC: CQ Press), 15–29.

NATIONAL GOVERNORS' ASSOCIATION (1990) *Educating America: State Strategies for Achieving the National Education Goals* (Washington, DC: National Governors' Association).

ODDEN, A. (1981) School finance; an example of redistributive policy at the state level, paper prepared for the School Finance Project, National Institute of Education.

ODDEN, A. and WOHLSTETTER, P. (1992) The role of agenda setting in the politics of school finance: 1970–1990, *Educational Policy*, 6(4), 355–376.

OGAWA, R. T. (1993) The institutional sources of educational reform: the case of school-based management, unpublished paper.

OLSON, L. (1993a, November 10) Novel voucher plan suffers resounding defeat in California, *Education Week*, 13(10), 1, 17.

OLSON, L. (1993b, November 17) Choice for the long haul, *Education Week*, 13(11), 27–29.

OLSON, L. (1993c, December 15) Who is afraid of OBE?, *Education Week*, 13(15), 25–27.

OSBORNE, D. (1988) *Laboratories of Democracy* (Boston: Harvard Business School Press).

PIPHO, C. (1979) *State Activity: Minimum Competency Testing* (Denver: Education Commission of the States).

PIPHO, C. (1986) Kappan special report: states move closer to reality, *Phi Delta Kappan*, 68(4), K1–K8.

PIPHO, C. (1989) Can the states agree to national performance goals?, *Phi Delta Kappan*, 71(4), 182–183.

PIPHO, C. (1990) Coming up: a decade of business involvement, *Phi Delta Kappan*, 71(8), 582–583.

PIPHO, C. (1993) Taxes, politics, and education, *Phi Delta Kappan*, 75(1), 6.

PLANK, D. N. (1988) Why school reform doesn't change schools: political and organizational perspectives, in W. L. Boyd and C. T. Kerchner (eds) *The Politics of Excellence and Choice in Education* (London: Falmer Press), 143–152.

PLANK, D. N. and ADAMS, D. (1989) Death, taxes, and school reform; educational policy change in comparative perspective, *Administrator's Notebook*, 33(1), 1–4.

PRESTINE, N. A. (1989) The struggle for control of teacher education, *Educational Evaluation and Policy Analysis*, 11(3), 285–300.

RAIMONDO, H. J. (1993) State budgeting in the nineties, in C. E. Van Horn (ed.) *The State of the States*, 2nd edn (Washington, DC: CQ Press), 31–50.

RAVITCH, D. (1990, January 10) Education in the 1980s, a concern for quality, *Education Week*, 9(2), 48, 33.

ROSENTHAL, A. (1990) *Governors and Legislatures: Contending Powers* (Washington, DC: CQ16 Press).

ROSENTHAL, A. (1993) *The Third House: Lobbyists and Lobbying in the States* (Washington, DC: CQ Press).

ROSENTHAL, A. and FUHRMAN, S. (1981) *Legislative Education Leadership in the States* (Washington, DC: Institute for Educational Leadership).

ROST, J. C. (1979) The times are changing in educational politics, *Thrust*, 8(5), 4–8.

RUDDER, C. F. (1991) Educational reform in Alabama: 1972–1989, in D. J. Vold and J. L. DeVitis (eds) *School Reform in the Deep South* (Tuscaloosa, AL: University of Alabama Press), 109–130.

RUDIAK, B. and PLANK, D. N. (1992) The politics of choice in Pennsylvania, paper presented to the annual meeting of the American Educational Research Association, San Francisco.

SACKEN, D. M. and MEDINA, M. Jr (1990) Investigating the context of state-level policy formation, *Educational Evaluation and Policy Analysis*, 12(4), 389–402.

SCHLAFLY, P. (1993) What's wrong with outcome-based education?, *The Phyllis Schlafly Report*, 26(10), 1–4.

SIMONDS, R. L. (1993) A plea for the children, *Educational Leadership*, 51(4), 12–15.

THOMAS, C. S. and HREBENAR, R. J. (1990) Interest groups in the states, in V. Gray, H. Jacob, and R. B. Albritton (eds) *Politics in the American States*, 5th edn (Glenview, IL: Scott, Foresman), 123–157.

THOMAS, C. S. and HREBENAR, R. J. (1991) Nationalizing of interest groups and lobbying in the states, in A. J. Cigler and B. A. Loomis (eds) *Interest Group Politics*, 3rd edn (Washington, DC: CQ Press), 63–80.

TIMAR, T. B. and KIRP, D. L. (1988) *Managing Educational Excellence* (London: Falmer Press).

TIMPANE, M. (1978) Some political aspects of accountability mandates, in E. K. Mosher and J. L. Wagoner, Jr (eds) *The Changing Politics of Education* (Berkeley, CA: McCutchan), 181–186.

TIMPANE, M. (1984) Business has rediscovered the public schools, *Phi Delta Kappan*, 65(6), 389–392.

TYACK, D. (1993) School governance in the United States: historical puzzles and anomalies, in J. Hannaway and M. Carnoy (eds) *Decentralization and School Improvement* (San Francisco: Jossey-Bass), 1–32.

USDAN, M. D., MINAR, D. W. and HURWITZ, E. Jr (1969) *Education and State Politics* (New York: Teachers College Press).

VAN HORN, C. E. (1993) The quiet revolution, in C. E. Van Horn (ed.) *The State of the States*, 2nd edn (Washington, DC: CQ Press), 1–14.

VOLD, D. J. and DEVITIS, J. L. (1991) Introduction, in D. J. Vold and J. L. DeVitis (eds) *School Reform in the Deep South* (Tuscaloosa, AL: University of Alabama Press), 109–130.

WALKER, R. (1990, March 7) With goals in place, focus shifts to setting strategy, *Education Week*, 9(24), 1, 17.

WEISMAN, J. (1991, September 18) In Indiana, business groups not talking as one on reform, *Education Week*, 11(3), 1, 18.

WIMPELBERG, R. K. and GINSBERG, R. (1989) The national commission approach to educational reform, in J. Hannaway and R. Crowson (eds) *The Politics of Reforming School Administration* (London: Falmer Press), 13–25.

WIRT, F. M. (1977) State policy culture and state decentralization, in J. Scribner (ed.) *Politics of Education* (Chicago: National Society for the Study of Education), 164–187.

5.   Politics of education at the federal level

Gerald E. Sroufe

Education is a state responsibility, a local function, and a federal concern. (US Secretary of Education, Richard C. Riley: March 11, 1994)

From a leading, and effective, spokesman of the centrality of education reform for achieving national economic and social welfare goals, this statement from the Secretary of Education envisions a notably modest role for the federal government in achieving these goals. To some extent, the Secretary's comment was offered as an antidote to the growing fears of state governors that the federal government was embarking on a program of increased mandates and centralization of education. It followed closely on the heels of a letter from Governor Carroll Campbell (R–SC) to the President decrying the escalating federal influence in historic state responsibilities. Nonetheless, if this is the most to be claimed for the federal role in education by the Clinton Administration's foremost advocate, one is required to consider the merits of devoting much attention to a topic of tertiary importance.

An examination of the contribution of the federal government on fiscal resources available to carry out elementary and secondary education bolsters the Secretary's minimalist perspective. While the federal share for elementary and secondary education declined by under 6% in the Reagan administration, it is unlikely to go beyond 7% even in the aggressively investment-oriented Clinton Administration. It has been remarkably consistent, varying only from 7.9% in 1966 to 8.1% in 1978 (Jennings 1981). Viewed as a percentage of federal expenditures, all education expenditures represent less than 2% of the federal budget. As noted by Verstegen (1987: 516), the federal government has had a role in the financing of education since the adoption of the Constitution, but it has never been a large role.

Given strong traditions of local control and clear state constitutional responsibility for education – and the magnitude of the reforms generated at the state level during the 1980s and 1990s in contrast with the modest efforts of the federal government – one must attend to the question of the relative importance of achieving greater understanding of the federal politics of education.

Why study federal politics of education?

1. *Study of federal politics of education illustrates and furthers conceptual analysis:* For instruction purposes, the federal politics of education offers many opportunities to illustrate, examine, and test various conceptual approaches to understanding the breadth and depth of political processes and phenomena. Because of the commonality of general experience and the plenitude of data, it is rewarding to teach politics of education using the federal experience. It is not accidental that the classical explication of concepts such as

0268–0939/94 $10·00 © 1994 Taylor & Francis Ltd.

Truman's (1951) 'group process approach,' or Schattschneider's (1960) 'scope of conflict' concept, depends almost entirely on federal examples. Conceptual or theoretical insight is no respecter of levels of government or political institutions, but it is easier to flesh out concepts with the rich data available at the federal level.

2. *Essential information about the federal politics of education is remarkably abundant and accessible*: The established procedures of the federal government make rich sources of information readily available. No state has publications equivalent to the *Congressional Record* or extensive legislative and conference reports, or routinely published reports of oversight, legislative, and appropriations hearings. Staff biographies, committee directories, member voting records, chronicles such as *Roll Call*, references such as *Politics In America*, electronic legislative reports, public access requirements of the Federal Advisory Committee Act (FACA), comprehensive coverage in the education press, numerous trade and popular books about legislation and politicians, and an extensive network of government relations specialists serve to provide the scholar with unrivaled access to information. Consequently, one good reason to study the federal politics of education, as noted by Eidenberg and Morey (1969) and Peterson (1974), is that it is possible to do so by heavy reliance on pertinent, generally accessible information of high reliability.

3. *The federal politics of education represents a singular set of institutions*: Scholars are attracted to study of the federal politics of education because of its singularity. There is essentially a common understanding of the federal government and its processes and procedures that is not shared among the states and local education systems. It is difficult for a single scholar to gain an equivalent understanding of several state political systems, and virtually impossible to generalize among the state and local systems. There is no need to seek generalizations about the federal politics of education as the system represents its universe.

4. *Because it is there*: Even though much less important in the lives of the typical student or teacher than the activities of their state legislature, the federal government is too large and noisy to ignore. The commotion created by federal politics of education is inescapable even while its motion is indiscernible. For example, the major piece of legislation of the Clinton Administration to date, Goals 2000, The Educate America Act, passed in 1994, has received nationwide media attention. Even if fully funded the reform legislation will provide extremely modest impact for extremely modest plans to support changes already under way in the majority of states. Still, most educators appear to be as well informed about this legislation as they are about the education politics of adjacent states.

For the reasons outlined above, it is reasonable to study the federal politics of education even when the policies and resources provided have little impact on schools. The argument advanced in this chapter should not, however, be understood as a plea for more attention to federal politics of education. Suggestions are offered below to stimulate a richer understanding of the federal system. Relatively speaking, however, priority should be given to improving the conceptual adequacy for study of local political systems, which seem mired in the community power structure studies of the 1960s (Hawley and Wirt 1974), and the state systems. An excellent start on both of these topics is made by colleagues writing adjacent chapters in this PEA Yearbook.

An important distinction: policy and politics

The Politics of Education Association (PEA) promotes the development and dissemination of research and debate on educational policy and politics. (Statement of Purpose, AERA Special Interest Group)

Most of our recent politics of education literature is actually devoted to education policy, a trend noted by William Boyd and Douglas Mitchell in their chapters in the *Handbook of Research on Educational Administration* edited by Norman Boyan. Writing in a section titled – ironically in the context of the present discussion – 'Politics and Policy,' Boyd (1988) notes the 'huge outpouring of policy studies and associated works' (p. 502). While Boyd offers a comprehensive critique of the limitations and contributions of policy studies, he does not seek to delineate between policy and politics. In his chapter, Mitchell (1988) calls attention to 'the shift in research on state systems from process-oriented political studies to more content-oriented policy analysis' (p. 458).

The thunderous movement toward prestigious policy analysis has, regrettably, been accompanied by a decline in political analysis. Despite the dualism as noted above in the purpose of the Politics of Education Association, we appear to have made a swap of policy for politics rather than having reached an improved synthesis. In making this trade, politics of education has mimicked developments within political science. Reflecting on his 'pilgrimage' as president of the American Political Science Association, Lowi (1992) stated:

> It is in this context [rational decision-making] that modern public policy became a hegemonic sub-discipline in political science, over-shadowing behaviorism itself. . . . The modern approach is more appropriately called public policy analysis, which draws upon macro-economic methods and economic system thinking. The best way to demonstrate the size and character of this new sub-discipline of political science is to point to the presence of policy analysis courses within the political science departments and the explosive growth of the separate policy analysis programs and the economic requirements in the schools of public affairs and public policy, and in the law schools. (p. 3)

The terms 'politics' and 'policy' have a common root, of course, and are often used interchangeably in everyday discourse. For the purposes of this chapter, and for general clarity within the field, and especially in addressing the 'politics of education' at the federal level, it is necessary to distinguish between the two terms.

Policy analysis and political analysis

Policy determinations are those which stipulate what is to happen. Policy analysis consists of examination of the degree to which what is happening is what is intended in an actual or preferred situation. Policy analysis generally involves evaluative questions: Are the intended children being served? Is the program cost-effective? What are the likely consequences of modifications in a program (e.g., national health care)? Of a new program? Policy analysis focuses on the question: What is the case?

Policies are a principal outcome of political processes, succinctly defined by Harold Lasswell (1936) as *Who Gets What, When, How*. The more formal definition, 'the authoritative allocation of values' (Easton 1965), helps explicate the 'what' of politics by introducing a broader range of goods, services, regulations, and symbols, among the 'values' allocated by the political system. Clearly, political analysis involves different questions than policy analysis. Rather than evaluate questions one considers aspects of the political system and its actors, processes, and structures in order to explain or predict who will get what. The goal of political analysis is to determine why what is of value is allocated in a particular manner or, most optimistically, what might be the case.

Political and policy analyses represent equally meritorious endeavors. They are often closely linked, as when policy analysis stimulates political actions necessary to provide new policy. For example, the Clinton Administration has sought to achieve a much higher concentration of funds available through Chapter 1 of the Elementary and Secondary Education Act (ESEA) for school districts with the highest concentrations of poor children. This policy initiative was created in large part from policy analyses indicating that many of the poorest children were not being served by existing policy which tended to allocate funds to all school districts.

Unfortunately, the policy suggested through these analyses did not fully address the political circumstances surrounding the issue. It soon became apparent that poor children in the districts of the chairman and ranking member of the House Education and Labor Committee, the chairman of the Subcommittee on Elementary, Secondary, and Vocational Education, and the president of the Senate would all lose funds for poor children in their districts and states through the redistribution suggested by the policy analysis. Congressman Dale Kildee, a leading supporter of a strengthened federal role in the education of the disadvantaged, noted that he wished to reach a policy understanding that 'would be effective, equitable, and would achieve 268 votes in the House.' Clearly, political analysis is different from policy analysis. The admirable education policy was scrapped in the House of Representatives because of the very predictable politics of the situation. Interestingly, one careful analysis of the implementation of Chapter 1 – a policy analysis – concluded that it had been ineffective in achieving the objectives sought largely because of the *political culture* of local school systems (Herrington and Orland 1991) as it responded to the objective policy mandates.

The importance of the distinction between politics and policy

The bulk of scholarly writing about the role of the federal government in education is devoted to analysis of policy rather than politics. One illustration of this singular focus, and the dysfunctions inherent in it, is provided through an examination of the policy analyses of the Reagan Administration provided by Clark and Astuto (1986). Perhaps the most provocative question before the Politics of Education Association for the past decade was kindled by their prolific publications arguing that the Reagan Administration had reversed fundamentally the trend toward an increasing federal role in education. The field is much indebted to the intellectual contributions of these scholars, which extend well beyond their extensive publications to leadership in many forums on campuses and in Washington. It is because they have written with such forcefulness and charity – and frequency – that their analytic posture is readily accessible for analysis.

One aspect of their analysis that is unequivocal is their exclusive consideration of policy at the expense of politics. For example, in 'The significance and permanence of changes in federal education policy,' (Clark and Astuto 1986) they use the term 'political' only once, and 'politics' not at all. They do use: policy, policy agenda, policy analysts, policy changes, policy continuum, policy debate, policy development, policy direction, policy interests, policy makers, policy preference (three times within six lines) education policy, social policy, and truncated policy.

There are two problems with the field's preoccupation with policy analysis. One is that it makes little contribution to the problem of explanation or prediction, which is generally considered a desirable aspects of scientific endeavors; a second is that it leads to arguable incorrect positions.

An important question which is beyond the range of policy analysis such as offered by Clark and Astuto, by way of illustration, is how the programs which the Reagan Administration sought to curtail came to be. Kaestle and Smith (1982) observed of the programs generated by the Johnson Administration 'that there has been a tremendous rise in federal aid to elementary and secondary education over the past twenty-five years, and that it has had a profound impact on local education' (p. 399). Similarly, Samuel Halperin (Halperin and Clark 1990) reports that in 1965 the Bureau of the Budget was unhappy with Title 1 of ESEA because it 'would limit spending to about $8 billion' (p. 20). There are political explanations for the profound federal education policy changes that occurred in the 1960s which are chronicled by Halperin and Clark and Kaestle and Smith (e.g., Graham's [1984] *The Uncertain Triumph*). While it is clearly a moral responsibility to pay attention to policy concerns (e.g., equity issues) at any point in time, it is equally imperative for scholars to seek to explain why such policies exist and under what circumstances they might be changed. Such analysis cannot to accomplished apart from considerations of the political system.

Exclusive attention to policy analysis, unfortunately, leads one toward static snapshots of phenomena that are more accurately represented as being in motion. Consequently, it was no doubt reasonable to observe, as did Clark (Halperin and Clark 1990) that 'In the last four years the National Education Association has been ineffectual in national politics . . . ridiculed by the President . . . helpless while the new Department of Education has been cut in half' (p. 24). But the day after the election of President Clinton, Ronald Brown went to the Association and declared, 'we could not have won the election without you,' and the NEA's pitiful fortunes improved considerably in the new administration not only with regard to education issues central to the organization (e.g., private school choice) but also on health issues vital to their members.

As is argued below, the Clark and Astuto analysis is vulnerable because it does not attend sufficiently to the political structures and process that help explain the policy phenomena identified, namely, the House Education and Labor Committee and the federal bureaucracy within and outside the Department of Education. The point, of course, is not that scholars such as Clark and Astuto do not know intimately the political aspects of the federal policies they decry, but that policy analysis leads one away from effective consideration of these essential political dimensions.

One seeking to apply the collective wisdom of the 'politics of education' to a practical task such as influencing an area of federal education policy is at a similar disadvantage to the policy analyst. In fact, even those who have written and taught courses about education policy are often poorly prepared to work within the political arena of the federal government. It is as though one has discovered most of a familiar board game: one knows in general how the game is supposed to go – what the objective is, for example – but if some of the key pieces and rules are not available, it is hard to appreciate the game fully or to play it well. It is also difficult to judge whether it is being well or poorly played.

Emphasis on policy matters provides steady work – there is literally no end to fresh policy considerations – but provides an incomplete understanding when, as often is the case, political matters are not sufficiently understood. Many of our efforts to chronicle policy events are unsatisfactory because they lack attention to the basic political structure and processes, thereby providing an incomplete and, frequently, erroneous understanding. Examples of additional pieces of the federal politics of education puzzle that must complement policy analysis are introduced below.

Additional pieces of federal politics of education puzzle

Politics does not permit policy development to end

For reasons of logic and convenience, most studies within the federal politics of education take a compelling piece of legislation (e.g., *The Uncertain Triumph* as a depiction of the politics of ESEA) or time frame (e.g., the Reagan years) as the unit of analysis. Generally, the narrative assumes that issues arise, are recognized and addressed by competing forces, and concluded. Occasionally epilogues are added to the episodic tales, as in Eric Redman's *Dance of Legislation* (1973), in an effort to accommodate the actual process from legislation to implementation. This artifact of traditional narrative style makes for clear and manageable presentation, but is very misleading. At the federal level issues are never fully ended; each period of quietude simply becomes the backdrop for the next stage of intense activity.

One familiar illustration of the continuing nature of policy development through implementation, of course, is Bailey and Mosher (1968). Their analysis uses the politics involved in the passage of ESEA legislation as but the preamble of policy decisions – and political struggles – associated with administration of the law.

Policy analysis seems particularly prone to analysis of events at a point in time, leading skilled observers to overstate the situation. For example, examining federal education policy at one point in time, Clark and Astuto (1986) assert that 'the current bipartisan consensus is a new consensus in support of a different federal role rather than a new consensus in favor of existing programs,' encouraging them to assert that the 'scope of redirection will be broadened over the next 5–15 years' (p. 4). Recent events demonstrate that bipartisan consensus has certainly gone from Congress, along with most of the Reagan Administration policy directives featured in the Clark and Astuto analysis. Indeed, given the historical perspective provided by the Bush and Clinton Administrations, it appeared that Lowi's (1992) observation about 'the inability of the Reagan Administration to terminate any important New Deal programs' (p. 6) also applies to its inability to eliminate the Johnson Administration's education programs.

There are several reasons why federal education policy issues never die, the chief having to do with politics. The beliefs and values of Sen. Edward Kennedy (D–MA) are simply not those of Sen. Nancy Kassebaum (R–KS), the ranking member on the Labor and Human Resources Committee during the 103rd Congress. Neither education advocate is likely simply to give up and move on to other issues, as do the weaker animals in the public broadcasting nature films. They will return to the issue over time and with regard to unrelated legislation; they will seek, and find, many other opportunities to express their values in education policy.

One opportunity to do so is in the annual appropriations process. One consequence of an annual appropriation is that few authorized programs in education are ever secure. For example, every US President since Dwight Eisenhower has sought to eliminate funds for impact aid (e.g., a federal contribution in recompense for assumed burdens of serving military dependents in the USA). None has succeeded, but the impact aid interests, including those in the bureaucracy, must always be in a state of readiness, and the political struggle is renewed each year.

A second opportunity to keep issues alive is through the authorization process. Education programs are typically authorized for five years. This means that there is a guaranteed review at that point, and that the contending parties are certain of a rematch.

Additionally, oversight hearings held frequently between authorizations give all parties an opportunity to search for advantage in achieving the values represented by an education program.

A third reason why issues are never fully resolved is that education is important as an issue of political symbolism. The key role of the National Education Association in the election of President Clinton assuredly has meant that there simply will be no more discussion of private school choice plans in his administration. Nonetheless, those advocating private school choice, such as Rep. Richard Armey (R–TX), find it in their interest to propose amendments to every piece of education legislation to secure their values, or at least to demonstrate to their supporters that they are working hard to do so. Even with zero chance of success, it is still in the interest of Rep. Armey to continue seeking to achieve his objectives, thereby preventing any issue from being closed.

Politics of the executive branch

An additional, important reason why the episodic treatment of education policy formation is not realistic is that a bill passed must still be administered: regulations drafted, staff hired, implementation policies created, and funds secured. Each of these tasks represents a political process very similar to the legislative process. The essential difference is that the process does not work to provide the same degree of access to the same groups, thus keeping issues 'resolved' by voting in the legislative process uncertain in the administrative process.

One illustration of the differential access provided by the administrative processes was the enforcement of civil rights legislation during the Johnson Administration. Achieving the Equal Educational Opportunity Program was regarded, appropriately, as a major victory for Johnson, attributed by many to his Southern roots and political expertise. However, little was achieved in the way of school desegregation under the program because of the political problems associated with its implementation. For example, the agency was constantly understaffed and funds were unavailable to fill allocated slots; regulations created a huge burden of proof on the federal agency; adjudication processes involved examiners provided by the Navy, traditionally the branch of the armed services that recruits most heavily in the South; and the administrative hearings were political mismatches: several low-level bureaucrats against a school superintendent and his Congressman and/or Senator.

This is not intended as an example of political hypocrisy. Rather, the example illustrates how the political forces that were insufficient to keep the desegregation legislation from passing were more than adequate to keep the policy from being effectively implemented. Federal political processes carry on beyond passage of the legislation, as was amply documented by Stephen Bailey and Edith Mosher in their seminal book, *ESEA: The Office of Education Administers A Law* (1968), and by Kirp and Jensen in *School Days, Rule Days* (1986) nearly two decades later. Obviously, some contenders have more access to the regulations and administration of programs than to the crafting of the legislation, but the latter has received most of the attention of scholars of the politics of education.

Few politics of education scholars have taken their research inside the executive branch. There are some excellent perspectives provided, however. One is Terrel Bell's (1988) chronicle of his years as US Secretary of Education seeking to work with President Reagan and his advisors. And no doubt someone will eventually explain why President Clinton offered only two paragraphs about education in his 1994 State of the Union address, one of which encouraged private management of schools. Clearly, there are winners and losers daily throughout the executive branch, and the political system associated with the administration of policy at the federal level is as worthy of attention as is the legislative system.

The centrality of congressional staff and committees

It is a natural tendency to associate federal education policy with paramount political figures. Newspapers and electronic media during the past 25 years have depicted President after President signing education bills. Hugh Graham (1984) demonstrates unequivocally how important individuals in the executive such as Samuel Halperin and Francis Keppel were to achieving the Johnson Administration education policies, as well as the importance of congressional leadership.

Periodically, a political scientist calls attention to the important role played by staff in the federal policy process. However, politics of education scholars seem not to have heeded these admonitions, preferring to chronicle the activities of those in most visible roles, such as the President. This is an unfortunate oversight, for to understand how any bill becomes a law requires an understanding of the staff dynamics associated with the bill. It is often stated that 'not every member of Congress voting on a bill knows everything that is in it.' Closer to the mark would be, 'no congressman knows everything in any bill, including those bearing their name.' Those who know, intimately, are the staff members and especially the members of the committee staff.

Among the important responsibilities of committee staff are to: (1) identify and secure witnesses for hearings; brief them; (2) prepare reports of hearings and conferences; (3) write language for legislation; (4) meet with interested parties inside and outside the legislature, including interest groups; (5) seek compromises with those opposed to the bill to gain support or at least inaction; (6) prepare a final conference report about the bill; and (7) brief members about the bill.

An illustration of the importance of staff can be viewed at any session of the House or Senate when legislation is being managed on the floor – the person handing the member the correct paper at the right time is a member of staff. During the Bush Administration legislation of the Office of Educational Research and Improvement was held over until the next session of Congress in part because a key Congressional staff member handling the bill resigned midyear. There was no way to proceed with the bill absent her expertise. A further case in point: The legislation signed by President Clinton in 1994 for the reauthorization of the federal research programs evolved over a four-year period; in the House 15 hearings were held involving nearly 100 witnesses. However, in the final sessions of the House and Senate conference on the bill only two staff members were involved, and they were the only two who knew clearly what final compromises were made and why.

An analysis of federal education research policy independent of federal politics of education research will be unlikely to offer any sense of predictability or explanation, and a satisfactory understanding of the federal politics of education requires much greater attention to the role and politics of staff in both the legislative and executive agencies.

Committees continue to be central

The modest reforms of the 1980s accomplished some diminution of the role of committee chairs, but did little to modify the centrality of the committee structure in shaping the federal policy process. As noted by Woodrow Wilson in his classic *Congressional Government* (1885/1981), the work of the Congress takes place in committees, and understanding the politics of education requires much more attention to committees than has been the case to date.

Congressional committees comprise political systems in their own right. They allocate hearing schedules, witnesses, field hearings, and other Congressional perks essential for re-election as well as influencing policy, on a political basis. The witnesses appearing on the third panel of a morning hearing are likely to be those recommended by the minority party; the field hearings are likely to be held in districts where memebers might be encouraged to adopt a more conciliatory view toward the majority members' values, or at least those of the chair of the committee.

Becoming a chairman or ranking member is the goal of every member of the Congress. Virtually all decisions regarding policy issues are designed with an eye toward committee assignments and leadership opportunities by members of Congress and their staff. The well-publicized political struggle within the Congress regarding the successor to the late William Natcher (D–KY) as head of the House Committee on Appropriations should not be viewed as an idiosyncratic event. Moreover, in this case, one individual who had been an active supporter of education and education research, Rep. David Obey (D–WI), successfully competed against a Representative who has not shown similar interest. While educators had little influence in determining the outcome of this internal political struggle, federal education policies will certainly be influenced by its outcome for years to come.

In recent years federal politics of education has resolved around the Democratic majority of the House Education and Labor Committee. The committee is easily the most liberal in the Congress. It includes strong representation from radical minorities and women, and has close ties with both the Black Caucus and the Hispanic Caucus. Issues of equity, which appeared to Clark and Astuto to have been a policy casualty of the Reagan Administration, have seldom been absent from the committee. In 1994, the debate over opportunity-to-learn standards (OTL) pitted the House committee against the Democratic administration, the parallel Senate committee, and the National Governors' Association. A reasonable understanding of federal education policy as reflected in Goals 2000, the Elementary and Secondary Education Act, and all other federal legislation will require investigation of the politics of the House committee.

The committee structure offers at least one other political consequence worthy of further examination. Not only does it serve to increase or reduce the influence of individuals (e.g., a strong education advocate such as Senator Robert Graham (D–FL) has little influence on federal education policy as he is not on the relevant committees), it moderates the influence of the waves of new members with new ideas that arrive with each national election. The freshman class, or the women's movement, tends to be dissipated as members are assigned junior positions on the various committees.

Time is the scarce resource

Dr Sally Kilgore (1988), speaking of her political learning from her years in Washington as

Director of Research within the Department of Education, proposes two concepts as central: time and territoriality. Her analogy, which is a useful one, asserts that all the actors were working on a two-year time line, and running about to put their individual flag on as many policies and programs as possible before their particular period as a federal policy maker expired.

Understanding the politics of education is facilitated by a sense of the time frames within which various actors operate. Despite what appears to be a leisurely work schedule, most policy is developed in an atmosphere of urgency, resulting in late-night drafting and even late-night floor debates. A familiar example is the recent passage of Goals 2000 in the Clinton Administration. The appropriations committees had provided FY 1994 funds for the program only if it were passed by April 1. The bill passed out of conference on March 18; and, following a cloture vote, passed the Senate at 1:30 am on March 26, just ahead of the deadline.

This type of time pressure has some deleterious consequences, especially as viewed from the scholarly, reflective perspective often most appropriate to policy analysis. Exploratory discourse is rare; ideas that do not lend themselves to succinct explanation, and clear and immediate policy implications have little prospect in the legislative arena. Similarly, ideas that cannot be reduced to 'language' (i.e., a sentence to be inserted or deleted at a particular point in a bill) have little prospect of becoming policy. Equally important, the timing of hearings and conferences is a major resource of the committee leadership in either promoting or sabotaging a legislative initiative. Important bills that are introduced late in the session become vulnerable to stalling tactics and are often simply deferred in the press to adjourn.

Individuals come with values

Our understanding of the politics of education would be enhanced by attending more to the values of the key actors. The formal models of policy analysis and political behavior tend to offer a highly rational model of behavior. The men and women who have held responsible positions in Washington over the past 25 years have acted from their experiences and values more than from formal policy analyses. Stephen Bailey (1950) devoted an entire chapter of his classic, *Congress Makes A Law*, to personalities, stating that attitudes and values of legislators were 'questions of cardinal importance to an understanding of the policy-making process' (p. 189). Bailey concluded his chapter of 'personalities' as follows:

> In a study of policy-making it is not enough that we understand influences external to the policy-maker. Constitutions and statutes, public opinion and pressures, facts and arguments, parties and patronage – these are factors which are important only as they reach and are interpreted and accepted by men's minds and prejudices. Like the action of light on variegated surfaces, external factors are absorbed, refracted, as reflected, according to the peculiar qualities of the minds they reach. (p. 218)

President Johnson's educational experiences, along with those of John Brademas, the first Greek-American to serve in Congress, offer more explanatory coins for the passage of the Elementary and Secondary Education Act of 1965 than many dollars' worth of demographic or economic studies. Rep. Major Owens (D–NY) is the first librarian to serve in Congress, and the school libraries and the Library of Congress have been the beneficiaries of his experiences. Sen. Tom Harkin (D–IA), who has a brother with a hearing disability, is a principal supporter of federal programs providing technical assistance for handicapped. Rep. William Goodling (R–PA), who was instrumental in the development of a federal program for preparing educational leaders, was a principal and a superintendent before entering Congress. The primary reason the leadership program was funded for three years on a declining basis is that this procedure reflects the core values of the Congressman: He could have received a larger amount of funds, for a longer period of time, without a built-in phase-out plan, had he chosen to do so.

As noted in Heineman *et al.* (1990):

> At its most basic level, policy analysis operates under the assumption that decision making ought to be a more rational process; analytic methods are assumed to enhance rationality in the policy process. Rationality does not describe very well how decisions are actually made, however. The very way in which humans think can lead to their values being more important in a decision-making situation than purely rational conclusions based on policy analysis. To understand policy making, one must understand policy-makers' values. (p. 56)

Assessment of the prospects for policy enactment, or full understanding of policy already enacted, requires consideration of the values of those involved in creating and implementing the policy.

Productive conceptual approaches to the study of federal education politics

General systems theory (Easton 1965, Wirt and Kirst 1992) has served well to introduce generations of new students to the idea of a political system. Its strength lies in its utility in organizing data and explaining political phenomena at a general level. It is likely the best and easiest teaching framework available. The most recent edition of the justly popular Wirt and Kirst (1992) text, *Schools in Conflict*, offers the field as good an exegesis of the political systems approach as we are likely to find. Unfortunately, the heuristic question begged by the authors, 'does it take us a small step in the right direction?' in their 1992 edition was first raised by David Easton in 1965. We have been taking very small steps indeed. Political systems analysis is ultimately not very satisfactory for examining the politics of education because politics largely occurs 'within the box' wherein system demands and supports are converted to outputs.

It may be more useful to consider an eclectic approach to investigation of the politics of education than to search for a unified theory, framework, or even general system. This is no doubt what Easton had in mind for political systems theory, his optimistic statement about heuristic research aside. It is clearly what Hans Morgenthau (Charleworth 1966) believed to be the wise course toward understanding politics:

> If one wants to make it [political science] into an exact science one has to despair, but if one wants to illuminate the political scene with theoretical insights, I do not believe one has to despair. (p. 148)

Some rich conceptual approaches have been provided in the classical literature of political science that have withstood the test of time, each serving to illuminate the scene of the federal politics of education.

David Truman's interest group approach

Historically, David Truman (1951) sought to move the discipline of political science from a normative understanding of 'pressure groups,' generally viewed as an aberration in the democratic system, toward an analytic posture. He concluded that not only were they normal, but essential to the American system of democracy. Truman coined the term 'access' as central to analysis of interest groups. While access and accessible are currently used to describe phenomena ranging from novelists to doorways, access continues to be a powerful concept for analysis of the behavior and influence of interest groups.

Strategy, procedures for enhancing access relative to one's rivals, and tactics and procedures for exploiting access were depicted fulsomely by Truman. With the possible exception of changes introduced by television and mass communications, there is little that any contemporary lobbyist could add to the concept developed by Truman. Private school choice is not a federal policy issue today, and the administration is not encouraging states to adopt private school choice plans, contrary to the fearful analysis of Clark and Astuto (1986: 11) just a few years ago, in large part because the National Education Association understands how to gain and exploit political access.

One particular tactic of note was explicated by E. E. Schattschneider in his small classic, *The Semisovereign People*. His concept of expanding or reducing the 'scope of the conflict' in order to change the relationships among political actors serves not only to explain political behavior but facilitates prediction and permits one to build political strategies. This is more payoff from a single concept borrowed from political science than can be gained from many more eloquent theories.

Morton Grodzins's American system

Morton Grodzins (1966) devoted his professional career to demonstrating the uniqueness of the American system of government in terms of mutually dependent levels of government. Though he grew to regret his analogy, he proposed a 'marble-cake' as opposed to the 'layer-cake' that was the food of traditional political science depictions of the federal system. He delighted in examples of sharing across levels, such as the local police force being trained at the national FBI academy, using equipment provided by the state government.

In describing this system of shared responsibilities and functions, Grodzins also sought to explain why it was the case. He concluded that it was because of the 'multiple cracks' in the system. He used 'crack' in two senses: (1) fissures or opportunities to have influence; and (2) wallop, to make a difference. One of the reasons it is so challenging to conduct analysis of federal education policies, as noted above, is that the process is never ending. A major reason why the process is never ending is that the structure and processes involved offer so many opportunities for policy to be made or shaped; there are many cracks in the system. Among the reasons for the many opportunities to influence policy within the American system are: the weak party structure, illustrated frequently by the low vote totals of the President compared with the high totals of many members of his party; the fact that there are separate appropriations and authorizing committees; that appropriations are required every year and authorizations, typically, every five years; that members of the House and Senate have different terms; the strength of the committees; and the procedures of the Senate which permit an individual to shape policy by engaging in delaying tactics.

An example of the impact of Senate procedures to provide points of access is the ability to introduce amendments on the floor. There were roughly 50 amendments introduced to the Goals 2000 bill, most of them offered by Senators not on the Labor and Human Resources Committee. Senator Charles Grassley (R–IA) and Sen. Jesse Helms (R–NC), for example, represented the religious right with successful amendments to the bill on the floor that they were unable to achieve through the Committee.

Murray Edelman's political symbolism

A final, useful classical text provides additional help in understanding federal politics of education. Educators, and even politicians, are apt to be discouraged by the policy outputs of an administration or Congress. Even Chairman William Ford referred to an education bill supported by his committee as a 'piss-ant bill.'

Murray Edelman's (1964) contribution is to declare that symbolic actions are as legitimate a function of the political system as substantive actions. Certainly, they are as important to politicians. We overlook much of the politics of education if we focus exclusively on substantive legislation, somewhat akin to dismissing the forest of Georgia pine because one knows there is a redwood growing somewhere.

Prospects for the PEA 50th Anniversary Yearbook

One hopes that over the coming decades the new, mostly young people who will shape the research paradigm of the politics of education will work toward recombining policy and politics into a new synthesis. The virtual abandonment of politics in favor of policy studies will have to be redressed. Additionally, the moral fervor of rationality must be moderated. In a chapter on 'Policy analysis and the political arena,' Heineman (1990) and his colleagues seem to speak for many educators as they express their displeasure with our circumstances:

> The policy analyst intent on bringing to fruition the findings of a carefully designed and executed study may discover too late that the traditional political preferences of Americans often create impassable political roadblocks. At the national level and in many state and local jurisdictions, vested interests are adept at exploiting institutional fragmentation and use their direct access to obfuscate and delay even the most rationally defensible plans. (pp. 113–114)

Margaret Mead (Dow 1991) made the same point, albeit more directly, commenting on the policy tangle encountered by the National Science Foundation resulting from its apolitical approach to curriculum planning, with regard to *Man: A Course of Study*: 'the trouble with you Cambridge intellectuals is that you have no political sense' (p. 206). In this instance, funding for education decreased from 42% of the NSF budget in 1960, to 2% in 1982.

When one reviews Heineman's concerns expressed in the paragraph quoted, it appears obvious that the policy/politics division of labor characteristic of politics of education research in recent years has run us aground. Clearly, the solution is to move our research paradigm from either policy dominance or political dominance to a synthesis methodology. Kirst is correct when he argues that we continue to work toward theoretically rich generalizations from experiences, and not retreat to simple description or outcome analysis. Sundquist (1968) provided such an analysis of ESEA, and other major legislative efforts culminating in the Johnson Administration. He traced the development of the issues over a decade, examining the problems addressed by the legislation, and the policy outcomes achieved. Finally, he sought to explain what the legislation achieved in terms of political factors, such as ideology, policy realignment, and congressional reforms.

Developing case studies at the federal level which seek to explain policy outcomes in the context of political phenomena, as illuminated by conceptual approaches, seems a good way to approach the next 25 years of study.

References

BAILEY, S. K. (1950) *Congress Makes a Law* (New York: Columbia University Press).

BAILEY, S. K. and MOSHER, E. K. (1968) *ESEA: The Office of Education Administers A Law* (Syracuse, NY: Syracuse University Press).

BELL, T. (1988) *The Thirteenth Man* (New York: Free Press).

BOYD, W. L. (1988) Policy analysis, educational policy, and management: through a glass darkly, in N. J. Boyan (ed.) *Handbook of Research on Educational Administration* (New York: Longman).

CHARLESWORTH, J. (ed.) (1966) *A Design for Political Science: Scope, Objectives, Methods* (Philadelphia: American Academy of Political and Social Science).

CLARK, D. L. and ASTUTO, T. A. (1986) The significance and performance of changes in federal education policy, *Educational Researcher*, October.

DOW, P. B. (1991) *Schoolhouse Politics: Lessons from the Sputnik Era* (Cambridge: Harvard University Press).

EASTON, D. (1965) *A System Analysis of Political Life* (New York: Wiley).

EDELMAN, M. (1964) *Symbolic Uses of Politics* (Champaign-Urbana: University of Illinois Press).

EIDENBERG, E. and MOREY, D. (1969) *An Act of Congress: The Legislative Process and the Making of Educational Policy* (New York: W. W. Norton).

GRAHAM, H. (1984) *The Uncertain Triumph: Federal Education Policy in the Kennedy and Johnson Years* (Raleigh, NC: University of North Carolina Press).

GRODZINS, M. (1966) *The American System* (Chicago: Rand McNally).

HALPERIN, S. and CLARK, D. (1990) Some historical and contemporary perspectives, in K. W. Thompson (ed.) *The Presidency and Education*, Vol. 1 (Lanham, MD: University Press of America).

HAWLEY, W. and WIRT, F. (eds) (1974) *The Search for Community Power* (Englewood Cliffs, NJ: Prentice Hall).

HEINEMAN, R., *et al.* (1990) *The World of the Policy Analyst: Rationality, Values, and Politics* (Chatham, NJ: Chatham House).

HERRINGTON, C. and ORLAND, M. (1991) Politics and Federal Aid to Urban School Districts: The Case of Chapter 1. In *Politics of Urban Education in the US* (Philadelphia, PA, Falmer Press).

JENNINGS, J. (1981) The federal role in paying for education in the 80s, in R. Miller (ed.) *The Federal Role in Education: New Directions for the Eighties* (Washington, DC: Institute for Educational Leadership).

KAESTLE, C. F. and SMITH, M. S. (1982) The federal role in elementary and secondary education, 1940–1980, *Harvard Education Review*, 52:4.

KILGORE, S. (1988) Symposium Comments, University Council for Educational Administration Conference, 'Reform of Administrator Preparation: What Universities Should Do.'

KIRP, D. L. and JENSEN, D. N. (eds) (1986) *Schools Days, Rule Days: The Legalization and Regulation of Education* (Philadelphia: Falmer Press).

LASSWELL, H. D. (1936) *Politics: Who Gets What, When, How* (New York: McGraw Hill).

LOWI, T. J. (1992) The state in political science: how we became what we study, *American Political Science Review*, 86(1).

MITCHELL, D. E. (1988) Educational politics and policy: the state level, in N. J. Boyan (ed.) *Handbook of Research on Educational Administration* (New York: Longman).

PETERSON, P. E. (1974) The politics of American education, in F. N. Kerlinger and J. Carroll (eds) *Review of Research in Education*, Vol. 2 (Itasca: F. E. Peacock).

REDMAN, E. (1973) *The Dance of Legislation* (New York: Simon and Schuster).

SCHATTSCHNEIDER, E. E. (1960) *The Semisovereign People* (New York: Holt, Rinehart & Winston).

SUNDQUIST, J. L. (1968) *Politics and Policy: The Eisenhower, Kennedy, and Johnson Years* (Washington, CD: Brookings Institution).

TRUMAN, D. B. (1951) *The Governmental Process* (New York: Knopf).

VERSTEGEN, D. A. (1987) Two hundred years of federalism: a perspective on national fiscal policy in education, *Journal of Education Finance*, 12, 516–548.

WILSON, W. (1885, 1981) *Congressional Government* (Baltimore: Johns Hopkins University Press).

WIRT, F., and KIRST, M. (1992) *Schools in Conflict: The Politics of Education*, 3rd edn (Berkeley, CA: McCutchan).

6. The international arena: the Global Village

Frances C. Fowler

The authors of the chapters about the other political arenas have an advantage over me; no one doubts that their arenas exist. However, the very concept of an international arena in the politics of education is contested (Ginsburg *et al.* 1990). Thus, my first task must be arguing that an international arena does indeed exist and that it can – and should – be studied. Having done that, I will use a brief literature review to indicate the state of the art of research on comparative educational politics and policy in the mid-1990s. Next, to illustrate some of the insights which can be gained from comparing the politics of education cross-nationally, I will discuss and analyze two international education reforms which have occurred in developed countries during the last 30 years. Finally, I will sketch a possible research agenda for the comparative politics of education. Although I believe that all countries participate in the international arena, for the sake of simplicity I have limited my discussion to the developed industrial democracies. These countries include the nations of Western Europe, the USA and Canada, Japan, Australia, and New Zealand.

Is the comparative politics of education possible?

The comparability of political systems

Researchers in the comparative politics of education encounter a common response to their work. Usually, at least one reviewer of each article and one member of each symposium audience question the fundamental validity of comparative studies. Objections typically take the form of asserting that political systems and the cultures which produce them differ so greatly that comparisons are impossible. This argument is normally supported by several myths. The researcher tries (tactfully, in most cases) to explode the myths and move on. However, those whose studies of the international political arena include the USA cannot move on quite as quickly as those who compare parliamentary systems, for a large body of political science literature argues that American politics is indeed 'exceptional.'

A thorough discussion of the 'exceptionality' literature would far exceed the scope of this chapter. In brief, it begins with the observation that, unlike other developed countries, the USA lacks both a strong labor movement and a powerful Labor or Socialist Party. Other frequently mentioned differences include the following: (1) American political institutions are based on the separation of powers while most other countries use a parliamentary system with fused powers; (2) American parties are undisciplined rather than disciplined; and, (3) American politics is not explicitly ideological (Coulter 1984, Theen and Wilson 1986). These are important differences which everyone who studies the international arena of the politics of education must bear in mind. American scholars, in

0268-0939/94 $10·00 © 1994 Taylor & Francis Ltd.

particular, must take care not to project the unique features of their own system onto other countries. Scholars of all nationalities need to maintain a keen awareness of the facts that even similar political systems differ in certain ways and that these differences grow out of underlying national cultures. However, important variations among institutions and cultures do not erase deeper similarities.

The deeper similarities

The advanced industrial democracies resemble each other in several ways. First, they are at similar points in the modernization process. All are urbanized and industrialized; all have literate populations; all have high standards of living; all are wrestling with the implications of the information age. Next, all combine some type of democratic political system with some type of capitalistic economic system. This means that all experience the inherent tension between democratic values – freedom, equality, and brotherhood – and the capitalist values of efficiency and economic growth (Swanson 1989). Finally, all participate in the world economy. Recession or expansion in one area affects other areas; global competition affects everyone. Inevitably, international economic forces influence national education politics. There is no escape; economically, all countries live in a global village (Wirt and Harman 1986).

The necessity for studying the international arena

Those who challenge the existence of the international arena implicitly accept the questionable assumption that national educational systems are 'largely autonomous, somewhat insulated, decision areas' (Merritt and Coombs, cited in Ginsburg et al. 1990: 477). The problem with this view is that it is not – and never has been – true. In the first place, it is ahistorical. No nation's educational system developed in splendid isolation from the rest of the world. On the contrary, from the beginning cross-national influences were important. For example, Sweden's school system was greatly influenced by Germany's. American ideas about public schooling were influenced by those of 18th-century French thinkers. Under the 19th-century Meiji Restoration, Japanese leaders toured the USA and Europe to learn how to set up a school system. Many such examples could be cited for every country.

Finally, the assumption of national autonomy ignores the presence of numerous education policy actors on the international stage. Many international organizations attempt to influence education policy. Among them could be mentioned multinational corporations like IBM and large foundations like Carnegie, Rockefeller, and Ford. UNESCO and OECD carry out cross-national research and disseminate it. Economic organizations like the International Monetary Fund and the World Bank also play major roles. Universities promote international encounters among scholars, leading to the exchange of ideas and information (Ginsburg et al. 1990). Increasingly, it is not possible to understand the education policy ideas which circulate in high places in one's own nation without also understanding what is happening in the international political arena.

The international arena: state of the art

As one might expect, given the contested nature of the international political arena and the many difficulties inherent in studying it, comparative research on educational politics and policy has developed slowly and sporadically. As a result, the literature is scattered and hard to locate.

The following literature review is based on a search of the *Current Index to Journals in Education* (CIJE) between 1969 and 1993, on an analysis of the reference sections of several major articles identified through the CIJE search, and on Harman's (1979) bibliography of research on the politics of education. It is, however, necessarily incomplete. The reasons for this incompleteness suggest some of the problems which beset researchers in this area of the politics of education. First, many contributions to the study of the international arena have appeared as chapters in edited books; such chapters are not included in standard indexes. Second, a number of contributions to the field have been published in 'minor' outlets such as the *Politics of Education Bulletin* – also not included in standard indexes. Moreover, I had access only to the indexes and catalogs available in a large university library in the USA. Although these references listed numerous sources in other English-speaking countries, they surely omitted many others that I could have located had I been able to make research trips to the UK or Australia – or even to Washington, DC. Moreover, the standard English-language indexes include few works in other languages, even in the important scholarly languages of French and German; and indexes of publications of these languages are not widely available in English-speaking countries. Finally, it is likely that sources relevant to the international politics of education are scattered through the political science, economics, sociology, public finance, religion, and legal literatures. Time and financial constraints made searches of all these literature bases impossible. Thus, the following review is of a limited literature. However, even this limited literature clearly reveals both the promise and the challenges of studying the international arena of the politics of education.

Two major streams of research appear in this literature. One – the earlier to develop – has been carried out largely by scholars in comparative education. The other, which was established in the 1970s and blossomed in the 1980s, is largely the work of scholars in educational administration who specialize in the politics of education. For the most part, these bodies of research have been developed by two groups of scholars who do not overlap and do not cite each other's work. Notable exceptions are provided by American Frederick Wirt and Australian Grant Harman, who publish in both literatures and often collaborate. Although I will discuss both bodies of literature, I will emphasize the latter. Not all works mentioned are cited in the reference section; however, enough information about them is given in the text to permit interested readers to locate them.

Comparative education came into its own in the 1950s; the major American journal in the field – *Comparative Education Review* – was founded in 1957. Its British counterpart, *Comparative Education*, was established eight years later. Scattered through their volumes and those of related journals are numerous articles about educational politics and policy. Most of these articles focus on the politics of educational reform; many also seem to be by-products of other research projects whose authors were so struck by the politics of a situation that they wrote an article about it. One exception is a series of studies sponsored in the mid-1970s by the Educational Policymaking in Industrialized Countries (EPIC) project and funded by the Ford Foundation, the Volkswagen Stiftung, and the University of Illinois (Wirt 1980). The comparative education literature includes research on a broad range of countries, both developed and developing. However, Western Europe, Australia, and the USA are better documented in it than are other countries.

The major strength of this literature is a major strength of comparative education generally; it is often well grounded in social theories and occasionally addresses methodological issues. The major weakness of this literature, however, is its unsystematic character. It would be difficult to use it to obtain detailed knowledge about the politics of education in any single country, even such major ones as the United Kingdom or Germany. Nor have culturally related groups of countries, such as the Nordic nations, been systematically studied and compared.

The literature produced by specialists in the politics of education developed later than the comparative education literature. It seems to have begun about 1970, when Wirt, Harman, and John Ewing published articles about the politics of education in the USA, Australia, and New Zealand respectively in the same issue of the Australian *Journal of Educational Administration*. Possibly it was their example which encouraged other scholars to explore comparative subjects, for in the mid-1970s scattered articles began to appear. For example, in 1973 Canadians John Bergen and Robert Lawson published separate pieces comparing Canada and West Germany. Two years later Australian Ian Birch published an article comparing states' rights in education in Australia, the USA and West Germany. Further impetus to comparative research was provided by the 1972 meeting of the Commonwealth Council for Education Administrations (CCEA), held in Fiji. American Jack Culbertson, then Executive Director of the University Council for Educational Administration (UCEA), delivered a paper at that meeting in which he proposed ways to promote international research. One result of his suggestions was CCEA and UCEA cooperation in the development of an international directory of scholars interested in doing comparative research on the politics of education. American Donald Layton was in charge of the project, and in 1977 and 1978 a directory and a supplement to it appeared (Layton 1977, 1978). During this same time period, Harman (1976) published an article in the *Politics of Education Bulletin*, calling for more comparative research and outlining a research agenda.

Another major step forward was taken in the late 1970s when the Center for Educational Research at Stanford and the Australian government began to fund the US–Australian Policy Project, directed by Wirt and Harman. In 1979 the project published a book by Jerome T. Murphy, comparing American and Australian education (Wirt 1980).

It could be said that the comparative politics of education came into its own in the mid-1980s. In 1986, Wirt and Harman published an important book on the subject, entitled *Education, Recession and the World Village: A Comparative Political Economy of Education*. In included an introductory chapter which sketched a theoretical framework for such studies and seven chapters about the effects of the recession of the early 1980s on educational policy in different countries. In the same year EAQ published a major article comparing the politics of the principalship in Australia and the USA (Chapman and Boyd 1986). After that, books and articles on various aspects of educational policy and politics in the international arena began to appear rather frequently in several countries. For example, the first Politics of Education Association Yearbook, published in 1987, included an introduction which firmly set American education reform in an international context (Boyd and Kerchner 1987). This volume of the yearbook was published in the UK, and simultaneously appeared as a special issue of the British *Journal of Education Policy*. Virtually every PEA yearbook since then has included at least one chapter about the international arena. Also in 1987, Boyd and Don Smart collaborated in editing a volume which compared Australian and American educational policy. It was followed in 1989 by a book edited by Boyd and James Cibulka about private school policy from an international perspective; in 1993 Hedley Beare and Boyd edited a book about school restructuring from an international perspective.

By the early 1990s articles on the international arena were appearing frequently in a broad range of journals familiar to the educational administration community. For instance, in 1990 EEPA published a comparative article about educational decentralization (Weiler 1990); and *Education Policy* (EP) published one about the United Kingdom's Education Reform Act. The next year, James Guthrie published a comparative article about the politicizing of educational evaluation in EEPA; EP offered three comparative articles, two about Israel and one about reform in developing countries. In 1992, EAQ carried an article about school choice in France, EP published a piece on the same subject as well as one comparing teachers' unions in France and Israel. Moreover, EP began to publish almost regularly on educational reform in the UK. Meanwhile, the British *Journal of Education Policy* was showing a similar interest in international themes.

Interest in the international policy arena is even spreading beyond the circle of scholars who specialize in politics and policy. For example, the American Education Finance Association's 1993 yearbook included *four* chapters on education reform abroad. Thus, as we move into the mid-1990s, this younger body of literature on the international politics of education is becoming increasingly rich and diverse.

This body of research, like the other, has both strengths and weaknesses. Its greatest strengths are its depth and its systematic character. Most scholars who write in this area focus on a small group of countries: the English-speaking nations and a few other nations which historically have been associated with them. As a result, a great deal is known about recent educational policy in the English-speaking world. Somewhat less is known about the political processes which surrounded the development of those policies, but even so this branch of the literature provides considerable information about educational politics as well. This literature is also relatively systematic. In it, scholars have addressed similar themes across several countries. For example, detailed descriptions of restructuring attempts or privatization reforms in many nations can be found in it. Thus, a good knowledge base exists, permitting in-depth comparisons of educational policy and politics within the English-speaking world.

Unfortunately, this focus on English-speaking countries can also be seen as a weakness. Although other nations have not been completely ignored, they have been less frequently and less thoroughly studied. As a result, while education policy in the USA, the UK, Australia, New Zealand, and Canada is well documented, even the major countries of Western Europe are only sketchily represented. The smaller countries of Western Europe rarely appear in this literature, and Japan has been almost totally overlooked (but see Sasamori 1993). A second weakness is the tendency of this literature to focus on describing political phenomena without trying to situate them explicitly in a theoretical framework. Although this atheoretical approach is perhaps natural in a literature which is still young, it must not continue indefinitely. A solid foundation has been laid, but much remains to be done.

Two international reform movements

Introduction

Arguing in 1976 for more international studies of the politics of education, Harman wrote:

Experiences in reasearch . . . in other countries help free us from the cultural and discipline restrictions of our own
environments. It helps us see the wood from the trees, and to be more critical of conventional ways of approaching
problems. (p. 1)

It is the purpose of this next section to demonstrate Harman's point by drawing on two
examples of international educational reform movements: the comprehensivization of
secondary schools in many countries since the Second World War and the efficiency
reforms of the 1980s and 1990s. In order to accomplish this purpose, a theoretical
framework will first be developed. Then each reform movement will be described and
analyzed. Finally, questions for further research will be suggested.

Theoretical framework

A major theoretical assumption of those who study the international arena is that the
world is interdependent and that global forces affect national education systems.
International economic developments are among the most important of these influences,
though others – such as wars and technological advances – also play a role (Boyd and
Kerchner 1987, Coombs 1984, Ginsburg *et al.* 1990, Wirt and Harman 1986). In addition,
cross-national evidence suggests that the relative importance of political values shifts
cyclically, at least among the industrialized democracies and perhaps more widely. From
roughly 1930 to 1980, educational policy in many countries focused on equity and social
justice. Since about 1980, an international shift to concern about freedom and excellence
has occurred, with a new emphasis on a cluster of values associated with economic
efficiency (Iannaccone 1987).

However, although nation-states are influenced by international forces, their
education policies are not mechanistically determined by them. Rather, global forces are
filtered through the 'prism of each country's unique characteristics' (Wirt and Harman
1986). These unique characteristics include economic resources, policy-making processes,
and national values. For example, global forces pressuring countries to develop more
accountability in education might stimulate several wealthy nations to invest heavily in
new testing programs. Relatively poor countries, however, might lack the funds to
undertake more testing. Their responses might be to develop less expensive procedures;
some might not even respond to this preesure.

A nation's governance and policy making structures also seem to influence its
reaction to international forces. Among the most relevant considerations in this area are:
(1) whether a parliamentary or a presidential political system is used and (2) whether the
country is a federation or has a unitary government (Wirt and Harman 1986). The type of
interest-group representation used also seems to be significant. In some countries, such as
the USA and the UK, many interest groups compete against each other informally and
lobby the government in a pluralist bargaining process. In others, such as Norway,
France, and Austria, the government and interest-group representatives meet formally to
negotiate in an approach to policy making called 'corporatism' (Duane *et al.* 1985, McLean
1988, Rust and Blakemore 1990).

Finally, each country has a somewhat different set of political values which include
ideas about how governments should act and what the purpose of education is. For
example, in some nations a 'contractual' political philosophy prevails, leading citizens to
understand the government as an impartial arbitrator which mostly acts to protect
individual rights. In such countries, people are likely to be suspicious of a government
which overtly assumes an active role in planning and directing education programs. In

contrast, there are other countries in which people hold an 'organic' political philosophy, believing that the government has a social responsibility 'to take action to construct and develop the economy and society in desired directions' (McLean 1988: 203). In them, citizens not only tolerate but expect the government to play an active role in planning and directing education programs. Obviously, the government of a country where a 'contractual' philosophy is dominant is likely to respond to international pressures by adopting education policies rather different from those which the government of a country with an 'organic' philosophy will prefer.

The following discussion will reveal that international forces have caused almost every country to attempt certain reforms. However, both the nature of those reforms and their relative success have been shaped by each nation's unique qualities, acting 'like a prism, refracting and adapting those influences, without blocking all of them' (Wirt and Harman 1986: 4).

Comprehensive school reforms

We tend to think of the 1980s and 1990s as the era of education reform *par excellence*, but education reform was high on the policy agenda in most advanced industrial democracies during the 1960s and 1970s as well. The specific reform that most countries attempted was comprehensivization, or the replacement of a highly differentiated, selective secondary education system with a more democratic and accessible structure.

In order to understand comprehensivization, one must first understand the educational status quo of much of the developed world in 1945. When European countries established their school systems in the 19th century, they chose a bi- or tri-partite, vertically scaled structure. France, for example, ran a bipartite school system. The *école* system, designed for the masses, provided a basic education up to age 14 for most children. The *lycée* system provided an academically oriented education for the children of the privileged classes; it alone offered access to universities. Other countries, such as Germany and Sweden, had similar systems which were tripartite rather than bipartite. The Japanese adopted the European approach after the Meiji Restoration of 1868. Thus, in 1945 almost all developed countries practiced segregation based on social class, restricting the best educational opportunities to the children of the middle and upper classes.

Several European leaders had expressed dissatisfaction with this system before the war, but change did not come until after it. The Japanese were the first to comprehensivize, though not by choice. Under the American Occupation, comprehensive education through grade 9 was imposed (Cummings 1982, Schoppa 1991). Sweden led the way in Western Europe; by 1971 the Swedish school system was comprehensive through grade 12 (Heidenheimer 1974, Husén 1989). Between the mid-1950s and the late 1970s, virtually all European countries attempted to replace their old, segregated systems with comprehensive ones, meeting with varying degrees of success. Today the school systems of Western Europe can be arranged on a continuum ranging from full comprehensivization (e.g., Sweden), to the more typical pattern of substantial comprehenivization (e.g., the UK and France), to a little comprehensivization (Germany), to no comprehensivization (Austria) (Boyle 1972, Budzinski 1986, Coombs 1978, Cooper 1989, Heidenheimer 1974, Hough 1984, Husén 1989, Judge 1979, Peterson 1973, Schwark and Wolf 1984).

These reforms represented a major change in educational policy for Japan and most of Western Europe. First, they involved a significant change in the structure of the school system. Additionally, they meant a shift in educational philosophy from a highly selective system designed to prepare a small élite to a less selective system designed to educate the masses.

Wirt and Harman's (1986) theory suggests that for the same policy change to occur in so many countries at about the same time, strong international forces must have been operative. Indeed, the evidence suggests that there were at least three. First, a major war had ended, establishing the USA as the unquestioned leader of the 'Free World.' The USA, of course, had a long history of comprehensive schooling (Cummings 1982). Next, the postwar period was a time of unparalleled economic growth in most developed countries; and rising standards of living led to rising educational aspirations. Finally, equity had been a major policy value in the international arena since the 1930s (Iannaccone 1987). All of these pressures apparently combined to encourage European countries to consider adopting comprehensive schools and to encourage Japan to keep them after the Occupation ended.

It is also relatively clear how the idea of comprehensive schools was diffused across national frontiers. International exchanges and conferences spread information about the American model. Soon there was a European model as well – Sweden. Many European educators travelled to Sweden to study comprehensive schools in action. Moreover, British political leaders invited several Swedes associated with the reform to visit the UK as consultants. Finally, an international policy actor, the OECD, adopted comprehensiviz-ation as one of its policy goals. It praised Sweden as an exemplary model, sponsoring international conferences which showcased the new Swedish system (Husén 1989). As late as 1979 the OECD pressured Austria to reform, issuing a report suggesting that Austria had a problem with equal educational opportunity (Budzinski 1986).

Yet, countries responded differently to these international influences. What sort of 'unique characteristics' contributed to these varying responses? The evidence suggests two which merit further exploration. It is striking that the countries which achieved full or substantial comprehensivization have unitary governments while the two which were least successful – Germany and Austria – have federal systems. This suggests that possibly it is easier to make major structural changes in education under unitary governments than under federal ones. It would be interesting to investigate this hypothesis by studying other structrural reforms cross-nationally. Second, it is hard not to wonder if Sweden (and other Nordic countries) succeeded so well at comprehensivization because of political values which place an unusually high priority on social equality. This hypothesis, too, could be tested cross-nationally by comparing education policies regarding other equity issues such as race, gender, and special education.

Efficiency reforms

Like the earlier comprehensivization reforms, the efficiency reforms of the 1980s and 1990s resulted from powerful global forces which affected all countries. The 1970s brought several rude awakenings to the developed world. In the postwar era, it had seemed that economic growth was natural, but two oil shocks, spiraling inflation, and mounting unemployment proved otherwise. Slowly people realized that a new set of economic conditions had developed which meant that rapid growth would not soon resume. In an effort to deal with what some Europeans labelled 'The Crisis,' political leaders around the world recommended policy changes, including educational reforms. These new reform proposals differed from the previous ones; the driving force behind them was not equal opportunity but efficiency. Politicians were concerned both about maximizing the yield from educational investments and about using the schools to develop a workforce which would enable their nation to compete effectively in the emerging global economy (Boyd and Kerchner 1987, Wirt and Harman 1986).

However, these strong global forces had to pass through the 'prism' of national characteristics just as comprehensivization had. Although some broad types of reform – such as changes in assessment practices and teacher education programs – were virtually universal, their precise nature varied considerably from country to country. Moreover, the preferred policy mechanisms for stimulating reform varied. Several scholars have noted that whereas the English-speaking countries have tended to favor market-oriented reforms such as competition, privatization, and school choice, these reforms have been less popular elsewhere (Fowler et al. 1993, McLean 1988, Rust and Blakemore 1990). In exploring this issue, I will provide some general information about the market-oriented reforms of the English-speaking countries and then describe different responses to similar policy proposals in Japan and France. Then I will suggest some of the questions raised by these differentiated responses.

The UK might be considered the leader in market-oriented education reforms. Under the Conservative government of Margaret Thatcher, the British implemented several reforms designed to increase competition and choice. One of the earliest was the Assisted Places Scheme, which since 1981 has provided voucher-like scholarships to relatively poor children so that they can attend élite private schools. In 1986, the government set up quasi-private City Technology Colleges, modelled on American magnet schools; they compete directly with the comprehensive high schools maintained by Local Education Authorities (LEAs). The culmination of Thatcher's market-oriented education reforms was reached in 1988, with the Education Reform Act. This law included several components designed to make British public schools more competitive. A national curriculum and national testing program made it easier for parents to 'comparison shop' for schools. Open enrollment in the public sector was permitted, and individual schools may 'opt out' of LEA control altogether. Thus, parents were given several ways to choose schools and to exert pressure on educators reluctant to meet their demands (Cooper 1989, Fowler et al. 1993, Guthrie and Pierce 1990, McLean 1988, Rust and Blakemore 1990).

Market-oriented reforms have not gone so far in the other English-speaking countries, but in all of them interest in or support for such reforms has been apparent. In the USA, Presidents Reagan and Bush actively supported public aid for private schools in the form of tuition tax credits and vouchers. Although some policies have not been widely adopted, other forms of competition have been, especially within the public sector. For example, selective magnet schools are common in large cities; and several states have adopted various forms of open enrollment within districts, between districts, and between high schools and universities. As the 1990s opened, 'charter' schools – similar to Britain's 'opted-out' schools – were being proposed. Moreover, corporations were beginning to develop plans to operate private schools or to manage public school systems (Boyd and Walberg 1990, Fowler et al. 1993).

Australia, Canada, and New Zealand have also been attracted to market-like reforms. In Australia, the political realities of the 1980s forced the Hawke Labor Government to abandon its plan to reduce public aid to élite, wealthy private schools. A coalition of private school parents and Catholic bishops convinced the Labor government that its electoral fortunes depended on continuing this aid, though it would permit Australia's loosely regulated private education sector to continue to weaken the government's own school system (Boyd 1989, Smart 1989). In the Canadian province of Ontario, the issue of extending public support to Catholic secondary schools (elementary schools were already supported) had lain dormant most of the time since 1928. However, in 1984 it reached the policy agenda again. In 1985–86 the Liberal government introduced and passed Bill 30, extending public aid to Catholic high schools. Challenged in the courts, the law was upheld in a 1987 decision (Lawton 1989). Like its English-speaking cousins, New Zealand has expanded the educational choices open to parents (Fowler *et al.*, 1993).

Japan and France provide illuminating contrasts with the English-speaking countries. Like them, both nations have been concerned about economic competitiveness and education reform. In both, ideas and issues similar to those discussed above have surfaced – but with different results.

In 1984 Japan's Prime Minister Nakasone recommended a set of education reforms to a group called the Ad Hoc Council. Among them were several which called for *jiyuka*, or 'liberalization.' The proposed *jiyuka* reforms should have a similar ring. Nakasone advocated reducing the power of the bureaucratic Ministry of Education and introducing 'private-sector vitality into the sphere of education' (Schoppa 1991: 224). Among the reforms which he believed would introduce vitality and competition into Japanese public education were establishing more private schools and making private 'cram' schools alternatives to regular high schools. *Jiyuka* was not a popular reform proposal, even within Nakasone's own Liberal Democratic Party. He also encountered predictable opposition in the Ministry of Education and *Nikkyoso*, the teachers' union. Moreover, public opinion was not on his side. As *jiyuka* moved through the series of consultative groups and meetings characteristic of Japanese policy making, it was criticized as being too 'individualistic.' Thus it was progressively watered down. By the time Nakasone's *jiyuka* proposal emerged fom the policy process in 1987, it had been reduced to a shadow of its old self. The final policy recommended 'study' of privatization and 'respect[ing] the wishes of parents' rather than school choice (Schoppa 1991: 235). In short, the reforms which had been so popular in the English-speaking world were quietly gutted in Japan.

Nor have they succeeded well in France. In fact, the French battle over government aid to private schools differs from Australia's in some illuminating ways. Like Australia, France has a large private sector which is generously financed by the government. Unlike Australians, however, the French heavily regulate private schools: the *explicit* purpose of the regulations is to protect the public system's 'preeminent' status. Moreover, the private system is not understood as a *competitor* of the public one; it is seen as a partner of the government in carrying out the educational goals of the nation. Nevertheless, throughout the 1960s and 1970s teachers' unions and the Socialist Party advocated nationalizing the private sector.

When Mitterand was elected president in 1981, the new government moved to make good on its promises; a bill was developed which in effect would have nationalized private education. After a massive demonstration in Paris in 1984, the government withdrew its bill. However, in 1985 it passed new regulations for private schools aimed – significantly – at making them more accountable for the public money which they received (Fowler 1992a, 1992b).

When the center-right came to power in 1993, it also moved quickly to make good on election promises – this time to expand aid to private education and relax regulations. However, in January 1994, parent associations, labor groups, and teachers' unions showed their opposition by organizing a demonstration of more than one million citizens in Paris. This government also backed down. Writing in *L'Enseignant* [Teacher] in the winter of 1994 about the demonstration, teacher union leaders put the government on notice that the French would not tolerate further attempts to privatize and 'Americanize' their public schools. Both French politicians and the French public seem more resistant to educational privatization proposals than their counterparts in the English-speaking world.

In light of Wirt and Harman's (1986) theory, one might ask several questions about the efficiency reforms of recent years. The global forces which encouraged them are clear in the literature, but the way that specific reform ideas spread across national boundaries is not. It would be interesting to go through the now abundant literature on these reforms, seeking to trace the development of the movement and to determine what specific international organizations, policy conferences, and scholarly exchanges played a major role in diffusing its ideas.

One might also ask several interesting questions about the differential responses of various nations to the movement. Among the English-speaking countries, for example, why has the UK been the leader in implementing market-oriented education reforms? Unlike the USA, Canada, and Australia, the UK has a unitary, parliamentary government. Is this yet another instance of the relative ease with which structural education reforms can be achieved under a unitary political system? Or should the explanation be sought in British political values? Or is it because the UK was subjected to an additional set of international pressures not experienced by the other countries . . . the unification of Europe? (For example, see McLean 1988.)

The cases of Japan and France are especially interesting because, like the UK, they have unitary, parliamentary governments. However, they differ from the UK in that they use a relatively corporatist approach to interest representation (Rust and Blakemore 1990). It is also likely that most citizens of Japan and France hold an 'organic' political philosophy rather than a 'contractual' one (McLean 1988). It has been suggested that both of these political characteristics are associated with a dislike for market-oriented education reforms. It would be interesting to know whether this theory holds true when applied to other countries. Moreover, what underlying values are suggested by the Japanese concern about the 'individualism' of the proposed reforms? Or by the French concern about 'Americanization'? Returning to the English-speaking world, it has been suggested that the Canadian province of Ontario has an interest representation system with corporatist elements (Duane et al. 1985). It might be enlightening, therefore, to explore Ontario's response to market-oriented reforms, comparing its response to the reactions of provinces and American states which use more pluralist approaches.

Each of these questions could provide the basis for several comparative studies, each with the potential for illuminating the current education reform movement and deepening our understanding of it. Yet, few of these issues would occur to scholars studying education reform in a single country. As Harman (1976) suggested, doing research on more than one country frees us from the limitations of our own culture and opens our eyes so that we can see new ways of conceptualizing the problems in our field.

Suggestions for further research

Among those who study the politics of education, research on the international arena is still relatively new. Fortunately, a good foundation for developing a full-blown research agenda has been laid. Scholars wisely began close to home, with studies of several English-speaking countries which are closely related to each other culturally. These efforts have established a solid, if limited, knowledge base and familiarized researchers with some of the methodological and theoretical issues which characterize work in this field. It is now time to build upon that base. In the next paragraphs I shall suggest what a research agenda for the next decade might look like.

First of all, a bibliography of research to date is badly needed. Such a project would have to be undertaken by a group of scholars including representatives from several countries. This bibliography should cover a wide range of types of publications and should be based on indexes and other compilations of works published not only in education but also in political science, economics, law, sociology, public finance, and religion. It should include developing countries as well as developed ones. Moreover, major works in languages besides English should be included, particularly in French, German, and Spanish.

Such a bibliography would make it easier for scholars who investigate the international arena of educational politics and policy to broaden the range of countries which they study. There are some signs that such a broadening has already begun. For example, articles on France, Germany, and The Netherlands are beginning to appear in the politics of education literature. This trend suggests that researchers are moving outward in a logical fashion from the English-speaking countries to the major nations of Western Europe. However, much remains to be done, even in Western Europe. France and Germany are still poorly understood; and, with the exception of France, the Latin countries of Southern Europe have not been studied. Nor have Belgium, Norway, Denmark, Switzerland, Austria, and Ireland been addressed. A matter of special concern should be the fact that Japan has been largely ignored thus far. As a major economic power which is also the only Asian example among the developed nations, Japan provides an interesting special case which could probably illuminate many educational policy issues in western countries. Once the advanced industrial democracies are better understood, the research agenda should be further expanded to include Eastern Europe and developing countries.

Study of the international political arena also needs to become more theoretical. It is, of course, difficult to know which theories are most relevant when developing a new line of research. However, the point has now been reached when those who do research on this political arena need to give serious attention to developing or deepening the theoretical components of their work. As the number of countries under study grows, it will become increasingly difficult to grasp subtle similarities and differences without situating those countries' political systems within strong theoretical frameworks. Theory-building should therefore be a major priority over the next decade.

Finally, in approaching all of the tasks outlined above, scholars in the politics of education should appropriate the comparative education branch of the literature. It includes studies of a wide range of countries which are poorly known among those who study the politics of education. It also includes several attempts to theorize comparative educational politics. A judicious appropriation of this literature would therefore accelerate the study of the international politics of education by providing added breadth and depth rather quickly.

It is an exciting time to be doing research on the international arena of educational politics and policy. A solid foundation has been laid, but much remains to be done. In a world which must increasingly be understood in global rather than national terms, such research promises to be both interesting and relevant.

References

BOYD, W. L. (1989) Balancing public and private schools: the Australian experience and American implications, in W. L. Boyd and J. G. Cibulka (eds) *Private Schools and Public Policy: International Perspectives* (London: Falmer Press), 171–192.

BOYD, W. L. and KERCHNER, C. T. (1987) Introduction and overview, in W. L. Boyd and C. T. Kerchner (eds) *The Politics of Excellence and Choise in Education* (Philadelphia: Falmer Press).

BOYD, W. L. and WALBERG, H. J. (1990) *Choice in Public Education* (Berkeley, CA: McCutchan).

BOYLE, E. (1972) The politics of secondary school reorganization, *Journal of Educational Administration and History*, 4(2): 28–38.

BUDZINSKI, E. (1986) Whatever happened to the comprehensive school movement in Austria? *Comparative Education*, 22: 283–295.

CHAPMAN, J. and BOYD, W. (1986) Decentralization, devolution, and the school principal, *Educational Administration Q arterly*, 22(4): 28–58.

COOMBS, F. S. (1978) The politics of educational change in France, *Comparative Education Review*, 22: 480–503.

COOMBS, P. H. (1984) *The World Crisis in Education* (New York: Oxford University Press).

COOPER, B. S. (1989) The politics of privatization, in W. L. Boyd and J. G. Cibulka (eds) *Private Schools and Public Policy* (London: Falmer Press), 245–268.

COULTER, E. M. (1984) *Principles of Politics and Government* (Boston: Allyn and Bacon).

CUMMINGS, W. K. (1982) The egalitarian transformation of postwar Japanese education, *Comparative Education Review*, 26: 16–35.

DUANE, E. A., TOWNSEND, R. G. and BRIDGELAND, W. M. (1985) Power patterns among Ontario and Michigan educational interest groups, in R. W. Common (ed.) *New Forces in Educational Policy-making in Canada* (Brock University of College of Education), 108–133.

FOWLER, F. C. (1992a) American theory and French practice: a theoretical rationale for regulating school choice, *Educational Administration Quarterly*, 28: 452–472.

FOWLER, F. C. (1992b) School choice policy in France: success and limitations, *Education Policy* 6: 429–443.

FOWLER, F. C., BOYD, W. L. and PLANK, D. N. (1993) International school reform: political considerations, in S. L. Jacobson and R. Berne (eds) *Reforming Education* (Thousand Oaks, CA: Corwin Press), 153–168.

GINSBURG, M. B., COOPER, S., RAGHU, R. and ZAGARRO, H. (1990) National and world-system explanations of educational reform, *Comparative Education Review*, 34: 474–499.

GUTHRIE, J. and PIERCE, L. (1990) International economy and national educational reform, *Oxford Review of Education*, 16: 179–205.

HARMAN, G. (1976) Politics of education research: an international perspective, *Politics of Education Bulletin*, 6(1): 1–3.

HARMAN, G. (1979) *Research in the Politics of Education, 1973-1978* (Canberra: Australian National University).

HEIDENHEIMER, A. J. (1974) The politics of educational reform: explaining different outcomes of school comprehensivization attempts in Sweden and West Germany, *Comparative Education Review*, 18: 388–410.

HOUGH, J. R. (1984) France, in J. R. Hough (ed.) *Educational Policy: An International Survey* (London: Croom Helm), 71–99.

HUSÉN, T. (189) The Swedish school reform – exemplary both ways, *Comparative Education*, 25: 345–355.

IANNACCONE, L. (1987) From equity to excellence, in W. L. Boyd and C. T. Kerchner (eds) *The Politics of Excellence and Choice in Education* (Philadelphia: Falmer Press), 49–65.

JUDGE, H. (1979) After the comprehensive revolution, *Oxford Review of Education*, 22(3): 480–503.

LAWTON, S. B. (1989) Public, private and separate schools in Ontario, in W. L. Boyd and J. B. Cibulka (eds) *Private Schools and Public Policy: International Perspectives* (London: Falmer Press), 171–192.

LAYTON, D. H. (1977) *The politics of education biographical directory* unpublished handbook.

LAYTON, D. H. (1978) *The politics of education biographical directory: supplement*, unpublished handbook.

McLEAN, M. (1988) The Conservative education policy in comparative perspective, *British Journal of Educational Studies*, 36: 200–217.

PETERSON, P. E. (1973) The politics of educational reform in England and the United States, *Comparative Education Review*, 17(2): 160–179.

RUST, V. D. and BLAKEMORE, K. (1990) Educational reform in Norway and in England and Wales, *Comparative Education Review*, 34: 500–522.

SASAMORI, T. (1993) Educational reform in Japan since 1984, in H. Beare and W. L. Boyd (eds) *Restructuring Schools: An International Perspective on the Movement to Transform the Control and Performance of Schools* (Washington, DC: Falmer Press), 136–156.

SCHOPPA, L. J. (1991) *Education Reform in Japan* (London: Routledge).

SCHWARK, W. and WOLF, A. (1984) West Germany, in J. R. Hough (ed.) *Educational Policy: An International Survey* (London: Croom Helm), 257–292.

SMART, D. (1989) The Hawke Labor Government and public–private school funding policies in Australia, 1983–1986, in W. L. Boyd and J. G. Cibulka (eds) *Private Schools and Public Policy: International Perspectives* (London: Falmer Press), 125–147.

SWANSON, A. D. (1989) Restructuring educational governance: a challenge of the 1990s, *Educational Administration Quarterly*, 25: 268–293.

THEEN, R. H. W., and WILSON, F. L. (1986) *Comparative Politics* (Englewood Cliffs, NJ: Prentice-Hall).

WEILER, H. (1990) Comparative perspectives on educational decentralization, *Educational Evaluation and Policy Analysis*, 12: 433–448.

WIRT, F. M. (1980) Comparing educational policies, *Comparative Education Review*, 24: 174–191.

WIRT, F. M. and HARMAN, G. (1986) *Education Recession and the World Village: A Comparative Political Economy of Education* (London: Falmer Press).

PART 3
Emerging research directions and strategies

7.   *Policy analysis and the study of the politics of education*

James G. Cibulka

Introduction

It is a frequent observation, particularly by European scholars whom I know, that the study of educational politics in the United States has been locked within an intellectual straightjacket, which is variously called pluralism, empiricism, pragmatism, behavioral science, or some other term meant to convey not a little derision.[1] Most scholars who work within the field of the politics of education probably would acknowledge that there is some truth in this characterization, at least until very recently.

Pluralism dominated the new field of study called 'the politics of education' to a much greater degree than it did in the parent discipline of political science. The earliest work in the field was pioneered by Thomas H. Eliot (1959), a political scientist with strong empiricist inclinations, and most of the work in the field for the next two decades followed this bent. The one textbook in wide use in the field, now in its third edition, authored by two political scientists Frederick Wirt and Michael Kirst (1992) who have made strong contributions to the field, retains a strong behavioral orientation.

Many who have contributed scholarship to the young field of educational politics and policy were trained within the field of educational administration rather than political science. For the most part they followed the cues of the parent discipline of political science at that time, albeit it lagging in their knowledge and application of developments in political science to the study of educational politics.[2] Correspondingly, the tradition of rationalism in organizational theory, and empiricism in scholarship, had a very strong foothold in the adjacent field of educational administration, and no doubt reinforced the pluralist biases these scholars brought to the young field of the politics of education.

Scholars in educational administration who specialize in the study of educational politics often have found themselves at odds with professors of educational administration, who were inclined to accept the admonition that politics and administration are (and ought to be) separate. Still, the terms of this debate were framed quite narrowly. The intellectual challenge which politics of education scholars posed to the field of educational administration was not rooted in fundamental disagreements about the value of empirical research or scientific knowledge, or about preferred political values. However, the political paradigm did challenge the rational-comprehensive model of organizational behavior. This same challenge eventually emerged within the field of educational administration itself. Critical theorists such as Greenfield (1986) and Foster (1986) pointed to the hidden role of values and interests in organizations. While this diagnosis displayed some affinity with the insights which politics of education scholars brought to an understanding of educational administration, most organizational theorists, even these critics, were not prepared to adopt a political model of administration, even one cleansed of positivist biases.

If this debate has lost some of its punch over the years, one suspects that it is for another reason: the politics of education as a field has undergone a major shift from a

0268–0939/94 $10·00 © 1994 Taylor & Francis Ltd.

behavioral paradigm to a policy paradigm. While this transition has been unannounced and rarely debated, it has been under way since roughly 1980. This paper deals with the nature of that shift and its implications for the study of the politics of education.

Organization of the paper

In the first section of the paper I will discuss the various streams of policy research and analysis. This will provide a framework for the second section, in which I discuss the contributions to policy research in the study of the politics of education, including the transition to a primary preoccupation with policy design in recent years. In the third and final section of the paper, after addressing earlier the problem of atheoreticism in policy studies, I will discuss two additional challenges facing policy research: the politicization of policy research and use of information in policy making.

Approaches to policy and policy studies

Definitions of public policy

It is helpful to begin with a definition of public policy. Dye (1992: 2) defines it as 'whatever governments choose to do or not to do.' Easton (1965) refers to policy very broadly as the 'authoritative allocation of values' for society. There are many other definitions, some focusing upon purpose, goals, means, and even practices. It other words, public policy can span a very broad terrain. It includes both official enactments of government and something as informal as 'practices'. Also, policy may be viewed as the inactions of government, not simply what the government does. In fact, these two examples alert us to how fuzzy the concept is. Some would deny that an informal practice of a government official is policy at all, even though it may have important implications for what governments do and who benefits or is harmed by such practices. Such practices may be administrative deviation from policy. More confusing still, the practices may be attempts by administrators to make sense of policy set by legislators or boards.

Consider also the problems which ensue when policy includes 'what governments choose not to do.' Typically, governments enact policy with an eye to solving some problem or ideal. Their failure to act may be intentional or may simply reflect a lack of awareness of a problem or need. Some would argue that this inattention has nothing to do with policy, while others would interpret it as reflecting hidden political processes, such as how powerful institutions shape language and our conceptions of problems and government. So the emergence of policy studies reflects many disagreements which have a long history in political science, disagreements which used to focus on how to define 'politics'. These same disputes now occur over how to define policy.

Definitions of policy studies

Policy studies are relatively new, spanning roughly the last two or at most three decades. The field has been a child growing up, in a manner of speaking, and as such has been struggling for a clear identity. There are many descriptors for the field – policy studies, policy science, policy research, policy analysis, and so on; indeed, the attempts to classify the aims of studying policy are almost as numerous as definitions of policy itself.

Most would agree that those who study policy have an interest in addressing social problems by generating policy-relevant knowledge. It is usually acknowledged that traditional preoccupations within academic disciplines such as political science and sociology have not addressed the amelioration of social problems directly, a defect many policy scholars are trying to rectify. For example, in political science, the behavioral concern with voting behavior, political parties, and public opinion has not shed much light on racial discrimination, poverty, and other pressing social issues. This problem orientation in the field gives rise to a common definition of policy analysis as 'the disciplined application of intellect to public problems' (Pal 1992).

Policy studies actually comprises a number of separate subfields with distinct, although overlapping, emphases. Accordingly, policy research draws broadly on many academic disciplines. Political scientists, sociologists, demographers, educational psychologists, evaluation specialists, and others contribute to the study of public policy. This diversity has contributed to the field's intellectual eclecticism, which is both a strength and weakness, as I will discuss later.

Table 1. Types of policy analysis.

Dimension	Basic (academic) policy analysis	Applied policy analysis
Aspect of policy which is of interest	Determinants, adoption, implementation, content, and impact	Content and impact
Aim	Understanding, explanation, and prediction	Evaluation, change, justification, and prediction
Actors	Researchers in universities or think-tanks	Consulting firms, interest groups, and government analysts

Table 1 classifies work in policy studies. Since there has been very good work in this area already, I have merely synthesized and elaborated on a classification by Pal (1992).

It is useful to view policy research as operating on a continuum from *basic (academic) policy analysis* to *applied policy analysis*. Academic policy analysis often has been called simply 'policy research.' Policy research, according to Weimer and Vining (1992), is closest to the traditional focus of academic social science disciplines, where the goal is theory development through specification of causal relationships and propositions grounded in empirical investigation. Policy research likewise focuses on relationships among variables describing behavior. It is distinct from much academic research in that it carries the narrower aim of analyzing social problems and the consequences of alternative government actions or inactions to address these problems. An early explication of this conception of policy research was offered by Coleman (1972).

Academic policy analysis (used synonymously here with research) can focus on a wide variety of phenomena related to the antecedents and causes of policy, how it is adopted and implemented, its content, and consequences. The aim of such endeavors is to understand and explain, and occasionally to predict. Typically, such analysis is conducted by researchers in universities or in academically oriented think-tanks.

By contrast, applied policy research, often referred to as policy analysis, concentrates on solving particular policy problems for policy makers or stake holders in the policy process. Those who practice it are often consultants, interest group representatives, or policy analysts in government who desire to evaluate, change or justify current policies, or on occasion to predict future consequences of existing policy or alternatives to it. Weimer and Vining (1992) describe applied policy research as 'analyzing and presenting alternatives available to political actors for solving public problems' (p. 4). This often leads to a client orientation which is very different from the way policy research is framed within academic disciplines.

Evaluation studies are a particular type of applied policy analysis. Some evaluation studies attempt to measure the outcomes or impact of policies through quasi-experimental or experimental designs, and in that sense are theory-based. At the same time, they generally contribute little information as to why a policy succeeds or fails, and therefore have limited utility to the social scientist who is attempting to build a theory of policy making. Process evaluation and formative evaluation do shed light on why certain processes are occurring, although these approaches cannot always link their insights back to outcome data. Some examples of evaluation research include studies of federal programs such as Head Start or the ESEA Chapter 1 compensatory-education program. Sometimes commissions issue reports which purport to 'evaluate' the adequacy of some aspect of American education. While often influential in shaping public opinion or the perceptions of policy makers, they do not satisfy the canons of rigorous evaluation research (Peterson 1983).

Two earlier reviews of educational policy research point to its many technical shortcomings (Boyd 1988, Mitchell 1984). Much of this research is more accurately described as 'applied' rather than 'basic' in its orientation. For example, Boyd (1988: 505) observes that many policy analyses focus on the benefits of some policy – often under consideration – while neglecting to consider its attendant costs.

Policy research in political science

Next I turn to a description of policy research which has been conducted in that discipline, so that we may assess linkages between the subfield of educational politics and the broader discipline. Table 2 describes four types of policy research which have been conducted in political science. It is adapted from Sabatier (1991). Much of the work in educational policy would fall under the first category, substantive area research. The focus here can be on

Table 2.   Types of basic (academic) policy research in political science.

Type	Examples
Substantive area research	Education, health, social welfare, crime
Evaluation and impact studies	Policy effect of environmental regulation
Policy process	Research on policy determinants, implementation, individual actor/public choice
Policy design/applied policy analysts	Alternative policy instruments, e.g., command and control versus economic incentives

Source: Adapted from Sabatier (1991).

health, transportation, crime, social welfare, housing, and so on. Evaluation and impact studies, which were just discussed briefly, constitute the second group. The third group of policy studies addresses the policy process, which includes study of the determinants of policies, how they are implemented, and other foci. Fourth, policy design research speaks to the efficacy of alternative policy instruments in relation to specific policy domains and problems within those domains. I have chosen to entitle this category slightly more broadly as *policy design/applied policy analysis* because much of the work in policy design addresses specific policy situations and problems, and in that sense can be incorporated within a broadened rubric.

These four categories, then, provide a useful framework for organizing the following section.

Overview of politics of education research with policy

The politics of education as an 'area study'

Political scientists who have assessed work in policy studies in the discipline have roundly criticized it for its atheoretical focus (Eulau 1977, Hoffebert 1986, Landau 1977). Part of this criticism has been that research is based heavily on the individual policy domain – its problems, unique context, institutional properties, and so on, with too little attention to common features which span different policy arenas. The heavy focus on descriptive case studies adds little to a cumulative knowledge base about politics across different policy areas. Using this argument, the study of educational politics and policy may contribute to understanding within its limited domain, but unless its research is framed theoretically, particularly within a comparative institutional context, research findings will not lead to cumulative knowledge about the broader political system.

Unfortunately, this criticism applies to the study of educational politics. Much of the work in the field has been atheoretical. While Wirt and Kirst's (1992) textbook attempts to synthesize the research in this field, their task is Herculean in scale. The disparate and idiosyncratic character of much politics of education research defies easy classification, much less building on confirmable generalizations.

At the same time, there is good research which is theoretically grounded. Consider Zeigler, Jennings, and Peak's (1974) study of the openness of educational policy making to citizen influence. The authors used representation theory and rigorous research methods. Even if their conclusions were controversial and open to criticism (Boyd 1976), the debate addressed theoretical issues important to an understanding of the political system. At the same time, this debate about how autonomous educational decision making really is was seldom linked to other policy areas (for an exception, see Tucker and Zeigler 1980). Indeed, in his assessment of this debate Peterson (1974) pointed out that the research evidence from other policy areas did not support the conclusion that educational policy making is particularly autonomous. Thus while the debate was framed in theoretical terms, its link back to the wider policy literature was inadequate. This may be a more accurate line of criticism about the better research conducted in this field – not so much that it lacks a theoretical base as that it still suffers from a provincial character. Comparative policy research is expensive to conduct, of course, but it would not be too much to ask that the broader political science literature inform research in this field more fully.

Evaluation and impact studies

The emergence of evaluation research in the field of education has been a growth industry since the 1960s, when federal dollars and policy leadership encouraged a more technocratic approach to policy development in states and local school districts. Many of the contributions to evaluation come from discipline-based scholars, but not particularly political scientists. The traditional preoccupation of political scientists with study of political behavior has produced a relatively small amount of evaluation research in that field.

Similarly, the large body of evaluation research in education, reflecting the size of this industry, has not been particularly informed by political science concepts and theories. There are exceptions. Consider Kirst and Jung's (1982) analysis of 13 years of Chapter 1 evaluation. But such evaluation efforts informed by political conceptions are rare. More typical is a study of Chapter 1 policy by evaluation specialists (Kennedy *et al.* 1987).

By contrast, the important debates about the schools' role in promoting equal educational opportunity and about the achievement effects of private schools have been dominated by sociologists, such as Coleman *et al.* (1966), and Coleman, Kilgore and Hoffer (1982). In part, this may be because of the methodological sophistication in quantitative methods associated with that discipline, despite criticisms of the morphological character of much early research on school effects. It may also reflect the relevance of sociological conceptions of social class, which has been shown to have such a pervasive impact on student achievement and, not surprisingly, on the structure of schooling and learning opportunities.

Recently, political scientists have begun to evaluate educational policy. Chubb and Moe's (1990) comparison of public and private high schools has caused as much flap in education circles as did Coleman and his associates in the early 1980s. Ironically, Chubb and Moe, relying largely on survey data, conclude that American public schools have become too political. Political scientists will not find a great deal of disciplinary knowledge at the heart of the Chubb and Moe analysis, although the book is indirectly informed by work in public choice theory, to which political scientists have made considerable contributions (See the chapter by Boyd *et al.* herein). Thus while the book represents a contribution to educational evaluation from a political science perspective, it does not tell us much about policy process. The book would have offered a more compelling critique of American schools had it drawn more heavily on process models to show how politics dominates the organization and administration of public schooling.

Chubb and Moe's book has been subject to a small mountain of criticism, some inappropriately amounting to personal attacks on the integrity of the researchers. Apart from Witte (1990) who has evaluated a Milwaukee private-school choice program, little other work in political science has evaluated education policy.

Because much of the research on educational politics is driven by the researcher's concern for improving the institution of schooling, this research has an evaluative thrust. In fact, the study of educational politics has been driven by changing policy concerns. In the early 1980s the management of declining enrollments became a research agenda. School officials appeared to be having difficulty managing decline because they ignored its political aspects. A considerable amount of work was done in this area by politics of education scholars (e.g., Boyd 1982, Cibulka 1987, Zerchykov 1981). More recently, the policy thrust toward coordinated services for children has been drawing politics and policy scholars' attention (Adler and Gardner 1994, Cibulka and Kritek in press, Crowson and Boyd 1993).

One must ask, however, how much these problem-driven issues contribute to a knowledge base about either politics or policy. Clearly, there is a need to draw together the disparate evaluation studies and to distill their meaning for the knowledge base in the field.

Perhaps the greatest impetus for change in the study of the politics of education has come from evaluation studies of the first wave of education reforms such as Timar and Kirp's (1988) comparative analysis of three states. These studies spawned an interest in designing new policies which are more effective (see below). Again, however, in most cases the contribution of this line of evaluation studies to a more general understanding of educational policy making has been limited.

Evaluation research does have the potential for making a greater contribution to educational policy analysis. Some of the recent advances within that field, particularly constructivist perspectives (Guba and Lincoln 1989) have implications for improving models of policy making. Implementers are seen as trying to make sense out of a complex and often bewildering set of circumstances. This has important parallels for recent research on the politics of policy implementation. In short, while it is unproductive to wall off evaluation studies from a broader understanding of policy, much research evaluating educational programs remains strangely disconnected from any inquiry into policy processes which create the observed impact. There is a desperate need for cross-fertilization among specialists in these domains. Unfortunately, disciplinary boundaries, self-defined specialties, and different orientations to the distinction between basic and applied research continue to be barriers to a broader approach to studying policy.

Research on policy processes

During the heyday of the behavioral revolution in political science, political scientists often ignored the content of public policies. The institutions through which these policies were administered also received little attention. However, some attention continued to be paid to the way policy processes operate.[3] Policy research began to receive renewed attention in the late 1960s and early 1970s, at the urging of Ranney (1968), Sharkansky (1970) and others. Some political scientists continued to work largely within a behavioral paradigm but welcomed the renewed emphasis on policy as an opportunity to understand policy processes better. Others saw this new policy focus as a way to apply knowledge to the improvement of political outcomes, using a more normative frame. Today, work on the policy process continues to reflect both 'basic' and 'applied' approaches. Consequently, work on policy processes has made one of the major contributions to policy research.

Beyond stage models and competing values frameworks

Stage models have been especially popular (Anderson et al. 1984, Jones 1977, Peters 1986). For example, consider Anderson's decision to study policies as they move sequentially from problem formulation, policy agenda, policy formulation, adoption, implementation, and evaluation. However, as Sabatier (1991) points out, these models are not really causal because they do not explain or predict how various stages of policy are linked. Therefore, they have limited utility for building a theory which links attributes of the policy process to outputs and outcomes. Further, the models tend to be inadequate in capturing environmental factors to which Easton's systems model does attend.

One of the most prominent examples of stage research is implementation theory, which is concerned with one stage in the total policy process. Educational researchers have contributed to the work on implementation theory which emerged beginning in the 1970s in political science (e.g., Bardach 1979, Pressman and Wildavsky 1986, Sabatier and Mazmanian 1979, Williams 1980, Williams and Elmore 1976). Implementation studies have occurred within the politics of education field and educational policy analysis (e.g., Berman and McLaughlin 1978, Elmore 1982, Kirst and Jung 1982, Peterson *et al*. 1986, Weatherly 1979, Wong 1990). In addition there is a broader literature on implementation in education that informs the politics of education, some of it in the literature on education reform (e.g., Odden and Marsh 1989).

This research illustrates the problems of stage theories. The early literature on implementation has been widely criticized as wrong in some of its key assumptions (e.g., Fox 1987, Palumbo and Calista 1990, Schwartz 1983). For example, much of the literature has drawn a sharp distinction between those who adopt policy and those who implement it. More recent research has clarified that those who implement policy, such as government agencies, contractors and even the target groups of the policy, also are intimately involved in policy design and evaluation. This fact makes it impossible to treat implementation as a conceptually distinct policy stage. Further, since policies frequently lack clear goals, it will not do to argue that implementors are the culprits who subvert policy. Rather, these implementors often reconcile design flaws and conflicting statutory objectives that many of the same people helped create at a policy's inception, a process of adaptation that makes the democratic process work better than it might otherwise. We now recognize that implementors have an explicit policy role, not merely a technical one.

While these insights gained after two decades of work might be regarded as legitimate advances toward a plausible theory of implementation, there continue to be two contradictory views of implementation, top-down and bottom-up. According to the former view, as described by Palumbo and Calista (1990), legislators specify goals in statutes and implementors set up regulations to ensure their fulfillment (Mazmanian and Sabatier 1981). The bottom-up view argues that 'street-level bureaucrats' (Lipsky 1980) such as teachers with their knowledge of what they and their clients (students) need, can adapt policies effectively through a process of 'backward mapping' (Elmore 1982). This leads to 'adaptive implementation' in which programs are institutionalized and responsive to program goals, although with some alteration in the goals of the policy (at least as some policy makers had envisioned them).[4] The constructivist perspective on program evaluation fits this latter conception of implementation.

It is not clear which view of implementation is 'correct,' and it may be that clarification of the utility of each perspective will require the development of models outside the policy implementation research. The development of a *comprehensive* or even middle-range theory of policy making might not treat implementation as a conceptually distinct stage. Certainly Easton was one of the first to encourage such a development. The 'feedback loops' in his systems theory were an attempt to capture the politics which occurs after policies are officially adopted, and how these shape subsequent 'demands' and 'support' flowing into the political system.

Alongside stage theory, there is another popular perspective among those who study educational politics and policy. This has been called a multiple-perspective approach (Knapp and Malen 1994). It is the idea that there are competing values in our political culture such as choice, equality, and efficiency (Guthrie *et al*. 1987). Similarly, Mitchell *et al*. (1989) employ quality, efficiency, equity, and choice.

For the most part, studies of educational politics use these values as heuristic devices rather than to build theory, in much the same way that stage theories have been employed. That is, research has established how such values are laden within various policies. To some extent the coalitions which support such values have been objects of attention, such as the role of business in advancing reform demands for greater efficiency and productivity. However, there has been little systematic examination of how political processes and structures facilitate or constrain the development of policies favoring one or more values. The problem is parallel to the reliance on stage theory. It is tempting to apply taxonomies such as stages or types of values to describe policies or stages of their development. This is far short of elucidating causal relationships among the relevant features of the problem system, however.

Furthermore, since values are environmental inputs to the political system, use of the value framework can treat the political system as a 'black box,' revealing little about processes and outcomes. Occasionally a study has attempted to take the classification a step further. Mitchell *et al.* (1989) link these values to seven policy domains in state policy making, using an analysis of laws, and then link these to differences in political culture among states. They employ policy, in other words, to study political culture as a dependent variable rather than to examine political process. The greatest promise for research would seem to rest in the use of values as an independent variable influencing political processes and outcomes (dependent variables). Nonetheless, the study represents a sophisticated attempt to use values, and the politics emanating from them, to model some aspects of political system behavior.

There is one further problem with the way values have been treated as a feature of educational policy by the educational politics literature. The 'competing-values' perspective assumes a value-neutral stance with respect to which values are preferred. It implicitly endorses a functionalist, equilibrium model and problems of system stability over system change. The closely related pluralist theory of counterbalancing, competing interests has been much criticized. Correspondingly, a limitation of the competing-values approach to understanding policy setting is that it offers no insight into why under-represented groups have difficulty securing a fair share of benefits through the political system. Critical theorists do offer various approaches to this problem (see, for example, the chapter on feminist and cultural studies perspectives by Marshall and Anderson, herein). Undoubtedly, there are a variety of ways to construct theories and models of policy making which capture power relationships more directly.

There is a need, therefore, to move beyond stage models and 'competing-values' perspectives on policy processes. In this section five of the major frameworks in use in political science and education, which have high potential for improving policy process research, will be reviewed. I have labelled these analytical frameworks as *systems theory*; *multiple decision-making models*; *policy types, streams, or arenas*; *institutional analysis*; and *critical theory*.

Systems theory: The work of David Easton (1965) and Easton and Dennis (1969) had a major impact in encouraging political scientists to move toward policy research, particularly the incorporation of policy outputs and impact into their work. Easton's influential systems analysis model provides the organizing framework for Wirt and Kirst's (1992) textbook on educational politics. The model did enjoy considerable saliency in the field of educational politics for a time, although sophisticated applications of the model linking environmental forces, system response, organizational 'throughputs' and policy outcomes generally have been lacking. The potential of the model for longitudinal purposes also has received less attention than it should, since the concept of feedback permits analysis of how and why policies evolve over time. One obstacle to the model's use is the heavy reliance on terms which can be prone to jargon.

Systems models can capture features of the larger cultural and socioeconomic environment which shape what enters the political system. Hoffebert's (1986) open system framework is a sophisticated attempt to model how these influences work. He posits that the socioeconomic environment and mass political behavior shape political processes.[5]

Multiple decision-making models: Peterson (1976) provided one of the most significant contributions to the use of policy models in the politics of education, drawing on Graham Allison's models explaining the Cuban missile crisis. Some models, for example, focus on the policy maker as a unitary actor, while others emphasize bargaining relationships among competing interests. Using policy cases from the Chicago public schools, e.g., desegregation and decentralization, he illustrates that particular models have greater explanatory power for some policy problem than others.

Accordingly, there may be no one best model to explain all kinds of policies. One criterion is the uses to which such models might be put. Policy is partly science, partly engineering and partly craft, and therefore it is framed differently by clients, professionals, and policy makers (Elmore and Associates 1990). Or the problem may be that the models need to be incorporated into a more comprehensive 'middle-range' theory (to borrow from Merton's classic terminology). The advantage of this approach is that it focuses model-building on particular features of the political system such as bargaining relationships or bureaucratic influences rather than attempting to capture everything in a systems model. While widely praised, Peterson's book has not influenced subsequent research as much as one might have hoped. This void reflects the lack of interest in theory-building among those who study educational politics and policy.

Policy types, streams or arenas: A third set of models tries to capture different slices of the policy process. Their grouping here is largely for convenience. What they share in common is their differentiation of the policy environment to understand how it works. The strategy can be to build a morphology of policy types, a classification of policy subsystems ('streams') focused on different issues and problems, or a coalition approach showing how policies get reshaped over time.

The earliest work in this tradition was done by Lowi (1964) and has proven a helpful organizing principle (e.g., Meier 1987), despite criticisms of the model's imprecision (Greenberg *et al.* 1977). In the area of education, Wong's (1992) attempt to explain different kinds of politics associated with different aspects of education policy making may be a useful point of departure for such inquiry.

Kingdon's (1984) work on policy streams attempts to model how policies get adopted. In a study of federal health and transportation policy, he documents the presence of three parallel, often independent processes: streams of problems, policies, and politics. New policy represents the convergence of fortuitous developments in these three arenas. While not yet applied to education, his work helps reframe the debate about the significance of ideas versus pressure and influence in political agenda-setting.

Sabatier and Jenkins-Smith (1993) offer an 'advocacy coalition framework' which posits policy change over time to reflect three sets of processes: the competition among competing advocacy coalitions (actors from a variety of public and private institutions with common beliefs and goals, and similar strategies) within a subsystem; changes external to the subsystem in the environment and other parts of the political system; and, finally, stable system parameters such as social structure and system rules. The latter two processes provide opportunities and constraints for the advocacy coalition. His work draws on elements of systems theory as well as a renewed focus on institutional behavior, discussed next.

Institutional research: The return to an interest in institutional behavior in recent years is a very eclectic development. In part this is because it draws on several disciplines, among them economics, sociology, and political science. In political science, for example, the so-called 'new institutionalism' differs from traditional theories of political institutions in critical ways, including basic assumptions (March and Olsen 1989).

A full description of the richness of the 'new institutionalism' is beyond the scope of this discussion (see Boyd, herein, for a fuller treatment). Two examples will illustrate its potential for informing the study of policy. One of its foci is how institutional rules affect behavior, and here public choice theorists have made seminal contributions. Public choice uses the individual actor as the starting point for identifying how institutional rules shape the behavior of organizational actors (March and Olson 1984). Keech *et al.* (1991: 219) have called this political economy approach 'choice-theoretic reasoning,' which has been applied by economists and political scientists in a variety of ways.

Kiser and Ostrom (1982: 180–181) describe the metatheoretical framework that uses a microinstitutional approach:

> It starts from the individual as the basic unit to explain and predict individual behavior and resulting aggregated outcomes. It is an 'institutional' approach because major explanatory variables include the set of institutional arrangements individuals use to affect the incentive systems of a social order and the impact of incentive systems on human behavior. Patterns of human action and the results that occur in interdependent choice-making situations are the phenomena to be explained using this approach.

Kiser and Ostrom's model posits that institutional rules, the attributes of the 'public good' being sought, and nature of the community (environment) shape the decision situation, to which the individual brings certain values and resources. In turn, the individual's actions, activities, and strategies lead to results. Policy is shaped at three levels – directly at the operational level and indirectly through collective choice arrangements (e.g., rules governing an agency), and at the constitutional level.

Not all proponents of the new institutionalism in political science adopt this reductionist and utilitarian public choice approach. For example, March and Olsen (1989) expound a perspective which attempts to explain the organizational basis of politics. This includes, among other things, explaining how meaning is elaborated through political institutions, a concern addressed by sociologists of organization. Policy pronouncements (as well as structures of decision making) are tools for enacting institutionalized myths in modern society, such as the salience of professional expertise in solving political problems (Meyer and Rowan 1977).

The new institutionalism has received relatively little attention until recently among scholars of educational policy and politics (for exceptions see Boyd 1992, Boyd and Hartman 1988), although it has received some attention in the literature on organizational theory (Mawhinney in press, Ogawa 1994, Powell and DiMaggio 1991).

Of course, the study of educational politics has leaned heavily on the traditional study of governmental institutions, such as the literature on intergovernmental relations. Yet those topics, too, are being influenced by newer approaches to the study of institutional behavior. For example, Vincent and Ostrom's model shows how the institutional rules at higher levels of government indirectly shape actions at the operational level.

Of particular interest to politics of education scholars are the literatures on federal policy and on state–local relationships. Work on intergovernmental relations has drawn on policy insights for the development of its theories, dating back to Grodzins (1966). Yet in much of the literature the dependent variable is government structure, not policy. For example, there is an unresolved debate over centralization versus decentralization (MacManus 1991). While some of the literature does take policy as the dependent variable, e.g., Wong's (1990) comparison of education and housing policies, it is the exception. This has led to rather separate streams of research which in fact have strong overlapping concerns.

Another example of this theoretical fragmentation is the literature on the impact of President Reagan's federal policies on state and local policies (e.g., Clark and Astuto 1986). Here the dependent variable *is* policy (state and local). However, the literature has been framed rather narrowly to address how the new federal policies affected state and local behavior, without much attention to how these impacts can help generate an improved model of intergovernmental relations. Intergovernmental relations is one example of how study of institutional behavior, if it is recast, can contribute to answering a larger puzzle: how to link policy problems/issues, political processes, and outcomes.

Critical theory: There is a considerable literature in education coming from what may be termed broadly a 'critical theory' perspective. This label really covers a wide range of theorists on the political left. Relatively little of this literature has a policy focus drawing on a political science tradition. Critical theories in educational administration tend to focus on organizational variables rather than the broader policy environment (e.g., Foster 1986, Greenfield 1986). Much of the work on micropolitics works from this perspective (Ball 1987, Blase 1991). (See the chapter by Malen, herein, for a fuller discussion of micropolitical models.) Sociological perspectives have had a heavy influence on critical theory, some of it with a policy emphasis (e.g., Popkewitz 1991). (For a more complete review of work in feminist and cultural studies, and its implications for the study of the politics of education, see the chapter herein by Marshall and Anderson, and the book by Marshall 1993). One of the few examples of a critical perspective (with a strong libertarian stream) employing a political and policy framework is by Spring (1993).

The distinguishing feature of this line of analysis is use of power to enforce social inequality of racial minorities, women, the poor, and underrepresented groups such as gays and homosexuals. While the breadth of such critiques defies easy summary, much of the work rejects a sharp distinction between political and social life, and focuses on hidden uses of power through such means as socialization and use of language. Such approaches to policy tend to reduce the focus on policy process as it has been described earlier, in favor of showing the importance of policy antecedents, e.g., inequitable distribution of wealth or access to information, or, alternatively, showing the link between policy outcomes and inequitable social outcomes, e.g., how unequal spending and resources among schools lead to school failure and attendant social problems.

It will remain a matter of debate whether policy processes are as insignificant in shaping policy outcomes and life chances as most critical theorists assume to be the case. One of the advantages of this approach is that it focuses explicitly on power arrangements, whereas many of the new policy models, including the new institutionalism, tend to lose this focus and thus are unable to explain how power really operates in a cumulative fashion. At the same time, many critical theorists tend to minimize the importance of societal values other than equality, such as efficiency and choice. Also, their analyses often can explain better how a social system preserves the status-quo rather than how that system changes and progresses. Some of these issues have concerned critical theorists working within the wider field of policy analysis (see the next section).

In sum, research on policy processes can be expected to proceed on a number of different fronts. In time, one or more of these lines of inquiry may gain ascendancy. But for now we can expect theoretical eclecticism.

Policy design

Most of the work in educational politics in the last decade has focused on education reform and the problems of designing policies to address the needs of reform. While this literature has developed largely independent of developments elsewhere in policy research there is, as Sabatier points out in his review of the literature, considerable work emerging in this area (Linder and Peters 1989, Salamon 1989). Some of this work is quite critical of the behavioral tradition in policy research (Bobrow and Dryzek 1987). This does not mean, however, that critical approaches eschew theory construction, as Forester's (1993) attempt at 'critical pragmatism' exemplifies. Pal (1992) also reflects a post-positivist orientation yet incorporates model-building. At the same time, other work continues to be done in a broadly positivist vein which is primarily oriented toward building a cumulative knowledge base (Schneider and Ingram 1990).

A major theme in the emerging policy design literature is how to select appropriate policy instruments which suit the policy problem (e.g., Cibulka 1990, Clune 1993, Elmore and Associates 1990, Fuhrman 1993, Malen and Fuhrman 1990, McDonnell and Elmore 1987, Schneider and Ingram 1990). This concern with the instrumental aspects of policy has tended to decrease emphasis on how such reforms reflect fundamental political processes, although such research has been conducted (e.g., Fuhrman 1988, Fuhrman and Elmore 1990, Heck and Benham in press; Malen and Ogawa 1988, O'Day and Smith 1992, Odden and Wohlstetter 1992, Smith and O'Day 1990, Timar and Kirp 1988).

The growing use of a policy design approach in study of educational policy has much merit if it can help make the study of educational policy and politics more relevant to the solution of educational problems. For instance, Crowson and Boyd's (1993) attempt to develop a model of effective coordinated services could be an important guide to future reform efforts in this direction.

Perhaps the most influential work in policy design is Smith and O'Day's (1990) 'systemic reform' model, which helped shape many of the education policies of the Clinton Administration. This framework is being used to revamp federal policy and realign state and local policy systems. This is a rare instance of concepts in the 'politics of education' actually reaching policy circles and becoming part of public policy. Systemic reform illustrates how influential the work in policy design can be. At the same time this theory has not been tested empirically. Rather, it grows out of the tradition of applied policy analysis. Policy design should not be seen as a substitute for the compelling need to build better theories and models of policy. Progress in that realm will not only advance basic research, but it should improve the quality of advice scholars can give to policy makers as well.

Future research challenges

Much of this review has pointed to the atheoretical tradition in policy research and the problems this has posed for development of the field. Two remaining problems are the politicization of policy research and the utilization of policy information to improve policy.

Politicization of policy analysis

Many political scientists have argued that policy analysis, far from being a scientific-technical enterprise, is itself a political resource used to advance partisan interests (Margolis 1974, Wildavsky and Tennebaum 1981). Much policy analysis claims that its focus is social improvement. But this raises a corresponding problem: the borderline between policy *research* and policy *argument* is razor thin. Policy argument is the use of research to fit a predetermined position with respect to a desirable policy or, alternatively, stretching findings to fit a predisposition. In the case of applied policy analysis, in contrast to basic policy research, the dangers of bias intruding into the work of those who serve clients seems obvious, since the clients are likely to color the way research is conducted or dictate its interpretation.

The biases of basic policy research are sometimes more subtle. Consider, for example, Hanushek's (1981) well-known critique of policies which pay teachers for advanced education and greater experience, and policies which lower class size. Hanushek employs sophisticated econometric tools to reach his findings, which are subject to numerous qualifications, but his conclusions interpreting these findings do not particularly display caution. For example, he characterized current spending policies as merely 'throwing money at schools,' a position which was eagerly embraced by the Reagan administration to undergird its fiscally conservative policies.

Coleman, Kilgore and Hoffer's (1982) study comparing public and private high schools, because of its controversial finding that Catholic high schools, *ceteris paribus*, produce higher achievement than public schools, quickly entered the arena of policy debate. Coleman and his associates were accused of engaging in policy argument by tailoring their analysis to permit the conclusion they wished to draw. These charges are malicious. However, the dispute does illustrate how policy research on controversial matters can be treated as if it were biased. The same scenario has surrounded Chubb and Moe's (1990) research comparing effective and ineffective schools.

The widening interest in education policy in recent decades arguably has politicized policy-oriented research in the field. More accurately, the latent political functions of policy analysis have become more widely recognized. On the positive side policy research can promote what Reich calls 'civic discovery' (1988: 3–4):

> The core responsibility of those who deal in public policy – elected officials, administrators, policy analysts – is not simply to discover as objectively as possible what people want for themselves and then to determine and implement the best means of satisfying these wants. It is also to provide the public with alternative visions of what is desirable and possible, to stimulate deliberation about them, provoke a reexamination of premises and values, and thus to broaden the range of potential responses and deepen society's understanding of itself.

Yet the proliferation of academic 'think tanks' in recent years illustrates how political resources influence what can be studied and how public opinion can be influenced and manipulated by deliberately slanted research. Thus, what has been called 'the argumentative turn in policy analysis' (Fischer and Forester 1993) can cut both ways, either as a device for enlightening the mass public and widening its influence over policy or, to the contrary, as a tool which works to legitimize the preferences of élites. There is a considerable body of research which portrays policy setting in the latter light (e.g., Dye and Zeigler 1975).

As an empirical fact, it will be debated for a long time what ultimate role policy analysts play in improving the effectiveness and responsiveness of our political system. As a matter of ethics, however, policy analysts must confront their obligation to capture and report the truth as they see it.

Policy information and analysis in policy making

This obligation in favor of truth-finding and truth-telling is greatly complicated by the fact that the most effective policy analysts in the political arena tend to shed the cloak of technical expert in favor of an advocacy role (Jenkins-Smith 1982).

Further, there is now a fairly conclusive body of research showing how information from policy research is used. Mitchell (1978) has studied this matter in education policy. Sabatier (1991: 270–271) summarizes the overall findings on the use of policy information as follows: within policy areas there is wide discussion of topics such as the severity of the problem, impact of past policies, consequences of alternative policies, and so on. However, such information tends to be used in an advocacy fashion by policy makers to do battle with opponents. The information is used to protect turf, delay action, or bolster credibility.

Still, policy information is more than a tool for short-term manipulation. Kingdon (1984) shows how timely policy information combines with perceived problems and opportune politics to establish new policy. Further, while individual studies rarely carry much weight, accumulated evidence can shape decision makers' perceptions. Policy research thus serves an enlightenment function (Weiss 1977) in which substantive policy information can play a role alongside ideology in shaping policy decisions. This influence of the policy analyst is accentuated by the fact that policy élites often form rather autonomous policy subsystems (e.g., Hamm 1983). Because the general public tends to have limited information and interest in many policy matters, policy élites do have considerable influence.

These overall findings about the role which policy information plays in policy making are ambiguous. They offer reassurance that there is a working space for the policy analyst wherein information is not entirely captive to partisan maneuvering. How to be politically effective, yet retain objectivity, remains a major tension, if not a dilemma.

Furthermore, insofar as policy analysis has made public policy-setting more complex, this poses a great challenge for democratic control. It is a specific manifestation of a fundamental problem in democratic theory and practice – the respective role which élites and masses play in the control of government and its policies.

How can we more effectively use data to change policy (Dery 1990)? More fundamentally, how can we design policy systems which use information more intelligently to improve their performance? Typically, we think of policy systems as means of regulating societal needs or resolving value conflicts. However, an equally powerful conception of policy systems sees them as learning systems much the same as organizations potentially are, capable of utilizing feedback and using this learning to improve their functioning (Argyris 1982, Sabatier and Jenkins-Smith 1993). The process may be more challenging in policy systems than in organizations, since the latter are more likely to have a unitary command structure. However, the differences are only matters of degree. There is a wide perception that our political system is no longer working effectively. This is at least in part a recognition of the limits of 'interest-group liberalism' as a system of governance.[6] Can policy analysis help us evolve a better democracy in which policy systems develop a greater capacity to improve themselves?

Conclusion

This review has raised issues both internal to the study of the politics of education and, more widely, about the role of policy analysis in improving education policy. A relatively narrow, but nonetheless vital issue, is whether we are witnessing the withering away of the study of educational politics *per se*, and its replacement by a broader field of inquiry which enjoys a similar label, 'educational policy studies.'

The study of educational policy and politics often is hyphenated or treated almost as though they are opposite sides of the same coin. Yet much of what is considered enlightenment in the study of educational policy has little or nothing to do with politics, as a cursory examination of what fills the pages of the journal *Educational Policy* will quickly reveal. The fact that many scholars of educational politics have now moved over to a policy paradigm has only added another layer of confusion to this development. For as I have taken pains to establish, the link between such policy research and the broader commitment to study of political processes has been sharply attenuated. Some of these problems of linkage between policy studies and political science research have beset the mainstream discipline. But the linkage has been all the more tenuous in educational politics research, which is one step removed from that discipline's knowledge base and therefore even less inclined to understand how the two might contribute to one another.[7]

While there is much to welcome as more policy research is conducted in education, its arrival will be lamentable indeed if it contributes to the demise of what we have learned in the study of educational politics since Eliot first charted an agenda for the field in 1958. Policy research should enrich that foundation and help it reach its potential, not commit it to a premature death.

I have argued that basic policy analysis can enrich the study of politics, while at the same time applied policy analysis can help us improve public policy making. Scholars in educational politics, drawing on the concepts and disciplinary knowledge of political science, have much to contribute to the discussions and debates about how we should redesign educational policies and educational policy systems. As specialists in the study of educational politics shift to more of a policy focus, they must take care that their identity as scholars with a special knowledge-base distinct from educational administration or other subfields is not diminished. If they heed this admonition, they will enrich educational policy studies and contribute to the maturing of policy analysis in political science itself. We may also hope that politics-of-education scholars will have some influence on the development of a new knowledge base in educational administration.

At the heart of policy analysis there will always be a tension between pursuing theory or practice, between capturing what is or seeking what ought to be. These are not new tensions or novel dilemmas. While policy analysis is a recent development, the role of experts and advisors in government is not. Many centuries before democratic government existed as we know it, emperors and kings were bedeviled by the problem of how to hold their experts accountable and use them to improve statecraft.

This old problem has taken on a new urgency today. Information and knowledge-based means of production are now a principal source of power, if not the primary factor shaping social and economic relationships (Castells 1989). Will this information revolution, which has helped to create the need for policy analysis in government, democratize the flow of information and increase the role of the citizenry? Alternatively, will it strengthen the ever narrower and autonomous spheres of experts possessing and controlling such information? The challenge facing policy analysis is to help us adapt democratic government to this new global transformation.

Notes

1. Students unfamiliar with these political science terms may wish to consult a basic political science textbook for details. For example, Lowi and Ginsberg (1992: 12) refer to pluralistic politics as 'competition among leaders or among powerful groups outside the government.' According to them (1992: 3) behavioral science 'focus[es] its attention primarily on the processes and behaviors associated with government . . . . [It does not] deal directly with the linkages between various processes and behaviors and the content of public policy.' Pluralism and behavioral political science have been criticized on many grounds, particularly their tendency to overstate the responsiveness of the political system to underrepresented groups and to ignore the unequal distribution of status, wealth, and power conferred by the political system.
2. One can always find exceptions to this caricature, to be sure. Political scientists working in the field may be least prone to such provincialism (e.g., Easton and Dennis 1969, Eulau 1972, Hawley 1975, Peterson 1976, Salisbury 1980, Wirt and Kirst 1992).
3. This focus never held entirely, of course. Lowi's (1964) distinction among distributive, redistributive, and regulatory politics generated considerable discussion and research in the field. Also, specialists in foreign policy emphasized its unique attributes, and many continued to pay close attention to the content of policies.
4. Some of the processes of adaptive implementation have been observed in federal education programs supporting educational change (Berman 1980, Weatherly 1979). The Elementary and Secondary Education Act, for instance, mandated both special aid to educationally disadvantaged children and general aid intended to placate diverse constituencies. This problem had to be resolved at the implementation stage and continuously reasserts itself during Congressional reauthorization proceedings.
5. Not all research supports this proposition. Also, the theory has been criticized and has undergone some revision. For a summary of the issue, see Sabatier (1991: 275).
6. Lowi (1969) first criticized this governance system as one controlled by powerful interest-groups pursuing narrow aims, in which the public interest is conceived as merely the byproduct of the equilibrium created by these interests. Interest-group liberalism is sometimes used synonymously with the term 'pluralism.'
7. Frederick Wirt has commented insightfully on this problem in a letter to the author. He argues that '[politics of education] studies are too much focused on the administration and evaluation of school policies. Leadership studies continue to be framed narrowly, political socialization is ignored, inter-governmental politics are underplayed, the role of the media is unexamined, the study of electoral politics has fallen dormant, there is little scholarly analysis of the courts, cross-national comparisons remain rare, and so on' (personal communication dated April 25, 1994).

References

ADLER, L. and GARDNER, S. (eds) (1994) *The Politics of Linking Schools and Social Services* (New York: Falmer).

ANDERSON, J. E, BRADY, D. W., BULLOCK, C. S. and STEWART, J., Jr (1984) *Public Policy and Politics in America*, 2nd edn (Monterey, CA: Brooks/Cole).

ARGYRIS, C. (1982) *Reasoning, Learning, and Action: Individual and Organization* (San Francisco: Jossey-Bass).

BALL, S. (1987) *The Micro-Politics of the School: Towards a Theory of School Organization* (London: Methuen).

BARDACH, E. (1979) *The Implementation Game* (Cambridge, MA: MIT Press).

BENHAM, M. K. P. and HECK, R. H. (1994) Political culture and policy in a state-controlled educational system, the case of educational politics in Hawai, i, *Educational Administrative Quarterly*, 30(4): 419–450.

BERMAN, P. (1980) Thinking about programmed and adaptive implementation: Matching strategies to situations, in H. Ingram and D. Mann (eds), *Why Policies Succeed or Fail* (Beverly Hills, CA: Sage).

BERMAN, P. and McLAUGHLIN, M. W. (1978) *Federal Programs Supporting Educational Change, Vol 8: Implementing and Sustaining Innovations* (Santa Monica, CA: Rand).

BLASE, J. (ed.) (1991) *Politics of Life in Schools* (Newbury Park, CA: Corwin Press).

BOBROW, D. and DRYZEK, J. (1987) *Policy Analysis by Design* (Pittsburgh: University of Pittsburgh Press).

BOYD, W. L. (1976) The public, the professionals, and educational policy-making: who governs?, *Teacher College Record* (May): 539–577.

BOYD, W. L. (1982) *School Governance in an Era of Retrenchment*, Final Report to The National Institute of Education (University Park, PA: Pennsylvania State University).

BOYD, W. L. (1988) Policy analysis, educational policy, and management: through a glass darkly?, in N. J. Boyan (ed.) *Handbook of Research on Educational Administration* (New York: Longman), 501–524.

BOYD, W. L. and HARTMAN, W. (1988) The politics of educational productivity, in D. Monk and J. Underwood (eds) *Microlevel School Finance: Issues and Implications for Policy* (Cambridge, MA: Ballinger), 271–308.

BOYD, W. L. (1992) The power of paradigms: reconceptualizing educational policy and management, *Educational Administration Quarterly*, 28(4): 504–528.

CASTELLS, M. (1989) *The Informational City: Information Technology, Economic Restructuring, and The Urban-Regional Process* (Oxford: Basil Blackwell).

CHUBB, J. E. and MOE, T. M. (1990) *Politics, Markets, and America's Schools* (Washington, DC: Brookings Institution).

CIBULKA, J. G. (1987) Theories of education budgeting: lessons from the management of decline, *Educational Administration Quarterly*, 23(1): 7–40.

CIBULKA, J. G. and KRITEK, W. J. (eds) (in press) *Coordination Among Schools, Families, and Communities: Prospects for Educational Reform* (Albany, NY: SUNY-Albany Press).

CIBULKA, J. G. (1990) Educational accountability reforms: performance information and political power, in S. Fuhrman and B. Malen (eds) *The Politics of Curriculum and Testing* (New York: Falmer), pp. 181–201.

CLARK, D. L. and ASTUTO, T. A. (1986) The significance and permanence of changes in federal education policy 1980–1986, *Educational Researcher*, (October) 4–13.

CLUNE, W. (1993) The best path to systemic educational policy: standard/centralized or differentiated/de-centralized, *Educational Evaluation and Policy Analysis*, 15(3): 233–254.

COLEMAN, J. S. (1972) *Policy Research in the Social Sciences* (New York: General Learning Press).

COLEMAN, J. S., CAMPBELL, E. O., HOBSON, C., McPARTLAND, J., MOOD, A., WEINFELD, F. and YORK, R. (1966) *Equality of Educational Opportunity* (Washington, DC: US Government Printing Office).

COLEMAN, J. S., KILGORE, S. and HOFFER, T. (1982) *High School Achievement: Public, Catholic, and Private Schools Compared* (New York: Basic Books).

CROWSON, R. and BOYD, W. L. (1993) Coordinated services for children: designing arks for storms and seas unknown, *American Journal of Education*, 1010(2): 140–179.

DERY, D. (1990) *Data and Policy Change: The Fragility of Data in the Policy Context* (Boston: Kluwer).

DYE, T. (1992) *Understanding Public Policy* 7th edn. (Englewood Cliffs, NJ: Prentice-Hall).

DYE, T. and ZEIGLER, L. H. (1975) *The Irony of Democracy*, 3rd edn (North Scituate, MA: Duxbury Press).

EASTON, D. (1965) A Systems Analysis of Political Life (New York: Wiley).

EASTON, D. and DENNIS, J. (1969) *Children in the Political System* (New York: McGraw-Hill).

ELIOT, T. H. (1959) Toward an understanding of public school politics, *American Political Science Review*, 52(52): 1032–1051.

ELMORE, R. F. (1982) Backward mapping: implementation research and policy decisions, in W. William (ed.), *Studying Implementation: Methodological and Administrative Issues* (Chatham, NJ: Chatham House), 18–35.

ELMORE, R. F. and ASSOCIATES (1990) *Restructuring Schools: The Next Generation of Educational Reform* (San Francisco: Jossey-Bass).

EULAU, H. (1972) Political science and education: the long view and the short, in M. W. Kirst (ed.) *State, School, and Politics: Research Directions* (Lexington, MA: Lexington Books), 1–10.

EULAU, H. (1977) The interventionist thesis, *American Journal of Political Science*, 21 (May): 419–423.

FISCHER, F. and FORESTER, J. (eds) (1993) *The Argumentative Turn in Policy Analysis and Planning* (Durham, NC: Duke University Press).

FORESTER, J. (1993) *Critical Theory, Public Policy, and Planning Practice: Toward a Critical Pragmatism* (Albany, NY: State University of New York Press).

FOSTER, W. (1986) *Paradigms and Promises: New Approaches to Educational Administration* (Buffalo, NY: Prometheus).

FOX, C. J. (1987) Biases in public policy implementation evaluation, *Policy Studies Review*, 7(1): 128–141.

FUHRMAN, S. (1988) State politics and education reform, in J. Hannaway and R. Crowson (eds) *The Politics of Reforming School Administration* (New York: Falmer Press), 61–75.

FUHRMAN, S. H. and ELMORE, R. F. (1990) Understanding local control in the wake of state education reform, *Educational Evaluation and Policy Analysis*, 12(1): 82–96.

FUHRMAN, S. H. (ed.) (1993) *Designing Coherent Education Policy* (San Francisco: Jossey-Bass).

GREENBERG, G., MILLER, J., MOHR, L. and VLADECK, B. (1977) Developing public policy theory perspectives from empirical research, *American Political Science Review*, 71 (Dec.): 1532–1543.

GREENFIELD, T. (1986) The decline and fall of science in educational administration, *Interchange*, 17 (2): 57–80.

GRODZINS, M. (1966) *The American System: A New View of Government in the United States* ed D. J. Elazar (Chicago: Rand McNally).

GUBA, E. G. and LINCOLN, Y. S. (1989) *Fourth Generation Evaluation* (Newbury Park, CA: Sage).

GUTHRIE, J., GARMS, W. and PIERCE, L. (1987) *School Finance: The Economics and Politics of Federalism* (Englewood Cliffs, NJ: Prentice-Hall).

HAMM, K. (1983) Patterns of influence among committees, agencies, and interest groups, *Legislative Studies Quarterly*, 8 (August): 379–426.

HANUSHEK, E. A. (1981) Throwing money at schools, *Journal of Policy Analysis and Management*, 1 (1): 19–41.

HARGROVE, E. (1975) *The Missing Link* (Washington, DC: Urban Institute).

HAWLEY, W. D. (1975) Dealing with organizational rigidity in public schools: a theoretical perspective, in F. Wirt (ed.) *The Polity of the School*, (Lexington, MA: Lexington Books), 187–210.

HOFFEBERT, R. (1986) Policy evaluation, democratic theory, and the division of scholarly labor, *Policy Studies Review*, 5 (Feb.): 308–329.

JENKINS-SMITH, H. (1982) Professional roles for policy analysts: a critical analysis, *Journal of Policy Analysis and Management*, 2 (Fall): 88–100.

JONES, C. (1977) *An Introduction to the Study of Public Policy*, 3rd edn (Belmont, CA: Wadsworth).

KEECH, W. R., BATES, R. H. and LANGE, P. (1991) Political economy within nations, in W. Grotty (ed.) *Political Science: Looking to the Future* (Evanston, IL: Northwestern University Press), 219–264.

KENNEDY, M. M., BIRMAN, B. F. and DEMALINE, R. E. (1987) *The Effectiveness of Chapter One Services* (Washington, DC: Office of Educational Research and Improvement, US Department of Education).

KINGDON, J. (1984) *Agendas, Alternatives, and Public Policies* (Boston: Little, Brown).

KIRST, M. W. and JUNG, R. (1982) The utility of a longitudinal approach in assessing implementation: a thirteen-year view of Title I, in W. Williams (ed.) *Studying Implementation: Methodological and Administrative Issues* (Chatham, NJ: Chatham House), 119–148.

KISER, J. and OSTROM, E. (1982) The three worlds of action, in E. Ostrom (ed.) *Strategies of Political Inquiry* (Beverley Hills: Sage), 179–222.

KNAPP, M. and MALEN, B. (1994) Assessing multiple-perspective analyses of educational policy making, paper presented at the Annual Meeting of the American Educational Research Association, New Orleans, LA.

LANDAU, M. (1977) The proper domain of policy analysis, *American Journal of Political Science*, 21 (May): 423–427.

LINDER, S. and PETERS, B. G. (1989) Instruments of government: perceptions and contexts, *Journal of Public Policy*, 9 (1): 35–58.

LIPSKY, M. (1980) *Street-level Bureaucracy: Dilemmas of the Individual in Public Service* (New York: Russell Sage Foundation).

LOWI, T. J. (1964) American business and public policy, case studies and political theory, *World Politics*, 16: 677–715.

LOWI, T. J. (1969) *The End of Liberalism: Ideology, Policy, and the Crisis of Public Authority* (New York: Norton).

LOWI, T. J. and GINSBERG, B. (1992) *American Government: Freedom and Power* (New York: W. W. Norton).

MACMANUS, S. A. (1991) 'Federalism and intergovernmental relations: the centralization versus decentralization debate continues, in W. Crotty (ed.) *Political Science: Looking to the Future Vol. 4: American Institutions* (Evanston, IL: Northwestern University Press), 203–254.

MALEN, B. and FUHRMAN, S. (eds) (1990) *The Politics of Curriculum and Testing* (New York: Falmer Press).

MALEN, B. and OGAWA, R. T. (1988) Decentralizing and democratizing the public schools: a viable approach to reform?, in S. B. Bacharach, (ed.) *Educational Reform: Making Sense of it all* (Boston: Allyn and Bacon), 103–120.

MARCH, J. and OLSEN, J. (1984) The new institutionalism: organizational factors in political life, *American Political Science Review* 78 (Sept.): 734–749.

MARCH, J. and OLSEN, J. (1989) *Rediscovering Institutions: The Organizational Basis of Politics* (New York: Free Press).

MARGOLIS, H. (1974) *Technical Advice on Policy Issues* (Beverly Hills, CA: Sage).

MARSHALL, C. (1993) *The New Politics of Race and Gender* (New York: Falmer Press).

MAWHINNEY, H. B. (1994) Institutional effects of strategic efforts at community enrichment, *Educational Administration Quarterly*, 31 (3): 324–341.

MAZMANIAN, D. A. and SABATIER, P. A. (1981) *Effective Policy Implementation* (Lexington, MA: Lexington Books).

McDONNELL, L. M. and ELMORE, R. F. (1987) Getting the job done: alternative policy instruments, *Educational Evaluation and Policy Analysis*, 9: 133–152.

MEIER, K. (1987) *Politics and the Bureaucracy*, 2nd edn (Monterey, CA; Brooks/Cole).

MEYER, J. W. and ROWAN, B. (1977) Institutionalized organizations: formal structure as myth and ceremony, *American Journal of Sociology*, 83: 340–363.

MITCHELL, D. E. (1978) *Shaping Legislative Decisions: Educational Policy and the Social Sciences* (Lexington, MA: Lexington Books).

MITCHELL, D. E. (1984) Educational policy analysis: the state of the art, *Educational Administration Quarterly* 20 (3): 129–160.

MITCHELL, D., WIRT, F. and MARSHALL, C. (1989) *Culture and Education Policy in the United States* (New York: Falmer).

O'DAY, J. A. and SMITH, M. S. (1992) Systemic school reform and educational opportunity, in S. Fuhrman (ed.) *Designing Coherent Education Policy* (San Francisco: Jossey-Bass), 250–312.

ODDEN, A. and WOHLSTETTER, P. (1992) The role of agenda setting in the politics of school finance, *Educational Policy* 6 (4): 355–376.

ODDEN, A. and MARSH, D. (1989) State education reform implementation: a framework for analysis, in J. Hannaway and R. Crowson (eds) *The Politics of Reforming School Administration* (New York: Falmer Press), 41–60.

OGAWA, R. T. (1994) The institutional sources of educational reform: the case of school-based management, *American Educational Research Journal*, 31 (3): 519–548.

PAL, L. A. (1992) *Public Policy Analysis: An Introduction* (Scarborough, Ontario: Nelson Canada).

PALUMBO, D. J. and CALISTA, D. J. (1990) *Implementation and the Policy Process: Opening up the Black Box* (New York: Greenwood Press).

PETERS, B. G. (1986) *American Public Policy: Promise and Performance* 2nd edn (Chatham, NJ: Chatham House).

PETERSON, P. E. (1974) The politics of American education, in F. Kerlinger and J. Carroll (eds) *Review of Research in Education*, Vol. 2 (Itasca, IL: Peacock), 348–389.

PETERSON, P. E. (1976) *School Politics: Chicago Style* (Chicago: University of Chicago Press).

PETERSON, P. E. (1983) Did the education commissions say anything?, *The Brookings Review* Winter: 3–11.

PETERSON, P. E., RABE, B. and WONG, K. K. (1986) *When Federalism Works* (Washington, DC: Brookings Institution).

POPKEWITZ, T. S. (1991) *A Political Sociology of Educational Reform* (New York: Teachers College Press).

POWELL, W. W. and DiMAGGIO, P. J. (eds) (1991) *The New Institutionalism in Organizational Analysis* (Chicago: University of Chicago Press).

PRESSMAN, J. and WILDAVSKY, A. (1986) *Implementation* (Berkeley: University of California Press).

RANNEY, A. (ed.) (1968) *Political Science and Public Policy* (Chicago: Markham).

REICH, R. B. (1988) Introduction, in R. B. Reich (ed.) *The Power of Public Ideas* (Cambridge, MA: Ballinger), 1–12.

SABATIER, P. A. (1991) Public policy: towards better theories of the policy process, in W. Grotty (ed.) *Political Science: Looking to the Future, Vol. 2: Comparative Politics, Policy, and International Relations* (Evanston, IL: Northwestern University Press), 265–292.

SABATIER, P. A. and JENKINS-SMITH, H. C. (1993) *Policy Change and Learning: An Advocacy Coalition Approach* (San Francisco: Westview).

SABATIER, P. A. and MAZMANIAN, D. A. (1979) The conditions of effective implementation: a guide to accomplishing policy objectives, *Policy Analysis* 5: 481–504.

SALAMON, L. (ed.) (1989) *Beyond Privatization: The Tools of Government* (Washington, DC: Urban Institute).

SALISBURY, R. (1980) *Citizen Participation in the Public Schools* (Lexington, MA: Lexington Books).

SCHNEIDER, A. and INGRAM, H. (1990) Behavioral assumptions of policy tools, *Journal of Politics*, 52 (May): 510–529.

SCHWARTZ, J. (1983) *America's Hidden Success: A Reassessment of Twenty Years of Public Policy* (New York: Norton).

SHARKANSKY, I. (ed.) (1970) *Policy Analysis in Political Science* (Chicago: Markham).

SMITH, M. S. and O'DAY, J. (1990) Systemic school reform, in S. H. Fuhrman and B. Malen (eds) *The Politics of Curriculum and Testing* (New York: Falmer Press), 233–267.

SPRING, J. (1993) *Conflict of Interests: The Politics of American Education* 2nd edn (New York: Longman).

TIMAR, T. and KIRP, D. (1988) *Managing Educational Excellence* (Philadelphia: Falmer Press).

TUCKER, H. J. and ZEIGLER, L. H. (1980) *Professionals Versus the Public: Attitudes, Communication and Response in School Districts* (New York: Longman).

WEATHERLY, R. A. (1979) *Reforming Special Education: Policy Implementation from State Level to Street Level* (Cambridge, MA: MIT Press).

WEIMER, D. L. and VINING, A. (1992) *Policy Analysis: Concepts and Practice*, 2nd edn (Englewood Cliffs, NJ: Prentice-Hall).

WEISS, C. (1977) Research for policy's sake: the enlightenment function of social research, *Policy Analysis*, 3 (Fall): 531–545.

WILDAVSKY, A. B. and TENNEBAUM, E. (1981) *The Politics of Mistrust* (Beverly Hills, CA: Sage).

WILLIAMS, W. (1980) *The Implementation Perspective: A Guide for Managing Social Service Delivery Systems* (Berkeley, CA: University of California Press).

WILLIAMS, W. and ELMORE, R. F. (eds) (1976) *Social Program Implementation* (New York: Academic Press).

WIRT, F. M. and KIRST, M. W. (1992) *The Politics of Education: Schools in Conflict* (Berkeley, CA: McCutchan).

WITTE, J. (1990) Understanding high school achievement: after a decade of research, do we have any confident policy recommendations?, paper delivered at the annual meeting of the American Political Science Association, San Francisco.

WONG, K. K. (1990) *City Choices: Education and Housing* (Albany, NY: State University of New York Press).

WONG, K. K. (1992) The politics of education as a field of study: an interpretive analysis, in J. G. Cibulka, R. J. Reed, and K. K. Wong (eds) *The Politics of Urban Education in the United States* (New York: Falmer Press), 3–43.

ZEIGLER, L. H., JENNINGS, M. K. and PEAK, G. W. (1974) *Governing American Schools: Political Interaction in Local School Districts* (North Scituate, MA: Duxbury Press).

ZERCHYKOV, R. (1981) *A Review of the Literature and an Annotated Bibliography on Managing Decline in School Systems* (Boston, MA: Institute for Responsive Education, Boston University).

8.   *Rational choice theory and the politics of education: promise and limitations*

William Lowe Boyd, Robert L. Crowson and Tyll van Geel

Few people would disagree that politics is a contest over 'who gets what, when, and how' (Lasswell 1936). Clearly and inevitably, it involves strategy and tactics, gamesmanship, bargaining, coalition building, and the like. How to 'win' at politics has fascinated humans since the dawn of history. But, until comparatively recently, the 'how to win' literature was mainly a collection of pithy insights and proverbs. It was far from anything resembling a science capable of analyzing, let alone predicting, the relevant strategic permutations. With the advent and development of rational choice theory – of which game theory (von Neumann and Morgenstern 1944), collective choice theory (Arrow 1951, Black 1958, Buchanan and Tullock 1962, Riker 1962), and economics are but three branches – all this has changed. Rational choice theory has altered the face not only of political science, but of sociology and organizational theory. While not dominant, and indeed often controversial in these fields (e.g., Lowi 1992), these new approaches, grouped under the rubric of rational choice theory, have permanently altered their landscapes. Moreover, beyond influencing the social science disciplines, the new approaches have had profound effects on the practical world of policy. The whole field of policy analysis (for better or worse) is heavily influenced by economic models and the cost–benefit paradigm. Government policies, big business strategies, and international calculations with respect to war and peace are increasingly influenced by mathematical modelling and game theoretic approaches (see, e.g., Bueno de Mesquita *et al.* 1985, Lewyn 1994).

Since the politics of education is a subset of the larger political scene – and a domain of growing strategic and fiscal importance in a competitive world economy – one might expect to see the best of the new analytical methods widely deployed to illuminate this arena. With a few notable exceptions (e.g., Peterson 1976, 1981), however, this is not the case.[1] Those specializing in the study of educational politics, most of whom are housed in colleges of education, have made only limited use of rational choice and economic models. Despite this, economic models nevertheless have had a profound impact on education policy (see Boyd 1992). The reasons for this state of affairs, the promise and limitations of these approaches, and the prospects for the future are the subject of this chapter.

This chapter is divided into three sections. The first section provides an overview of rational choice theory and its various branches. Section two reviews a small amount of work that has relied on the rational choise paradigm to study the politics of education. The third and concluding section comments on the promise and limitations of this approach.

0268–0939/94 $10·00 © 1994 Taylor & Francis Ltd.

The advent of the 'new' analytical approaches: theory and fundamental findings

What are called 'new' approaches here are not really new, as they date from the 1940s and 1950s. But they are foreign enough to traditional ways of thinking about politics (from political science, sociology, and social psychology) that they have been slow to catch on widely. Like many things 'foreign', some view them as suspect and maybe even harmful. Economics, of course, is widely known as the 'dismal science,' a discipline bereft of a human heart, insensitive to inequality and human suffering. It is a field, that, many feel, should be condemned rather than emulated. But it is easier to condemn than to stamp out economic behavior. Similarly, many people view with horror politics and the manipulative gamesmanship that goes with it: These are reprehensible things to be avoided, if they can't be eliminated. Sadly, for those of this persuasion, politics not only cannot be eliminated, it is impossible to avoid. Political behavior and gamesmanship occur even in organizations that strive to minimize hierarchy and maximize democracy in the workplace.

Few will disagree that democracy is a laudable goal. Democracy based on majority rule, and coupled with adequate legal protections for individual and minority rights, provides a framework for fairness and social justice. But what if there are problems in devising voting schemes that consistently and reliably reflect majority preferences? Unfortunately, there *are* such problems and astute politicians have recognized and exploited this fact (Riker 1986).[1] One of the greatest contributions of rational choice theory has been to document and analyze the very significant ramifications of these problems. Many scholars who embrace democracy and decry the individualistic, 'self-interested' approach of rational choice theory fail to appreciate the inescapable nature of these problems, to which we now turn.

Broadly speaking, to offer a rational choice of social phenomena means nothing more than saying that social phenomena result from individual human actors 'whose actions are directed by their beliefs, goals, meanings, values, prohibitions, and scruples. Human beings, that is, are *intentional* creatures who act on the basis of reasons' (Little 1991: 39). Based on this assumption, a historian, sociologist, political scientist, economist and even an anthropologist can seek to offer explanations of a social phenomenon, e.g., crime rates. Thus, in a real sense rational choice theory is not new at all but dates from at least the origins of microeconomic theory, with its theories about how individuals and firms make choices, taking into account the costs and benefits of the alternatives they confront (Alchian and Allen 1977). It is even argued today that Marxism is, at bottom, an elaborate rational choice paradigm (Little 1991).

What is new is the effort to use the basic starting assumption systematically to build models of social and political life. The branch of rational choice theory known as game theory had its origins in the 1940s and 1950s, the same period in which the theory of collective choice was being developed by Duncan Black (1948, 1958), Kenneth Arrow (1951) and Anthony Downs (1957).

Many people find this way of looking at human behavior in general, and politics in particular, uncongenial and alien. The sense of strangeness some people feel toward rational choice theories is perhaps best addressed by outlining the general idea of the approach and then elaborating on some of the branches of the field. Rational choice theorists attempt to make sense of the complexity of human action by starting with a few simple, but ultimately powerful, assumptions or axioms about human motives and behaviors. By deduction, they develop the logical implications of the assumptions to explain and predict behavior, taking into account the behavior of others and different 'institutional' settings, such as voting rules, agenda-setting rules, and incentive systems. As a social science perspective resting on 'methodological individualism,' rational choice theory claims that all social phenomena are derivable from the properties of individuals, after taking into account the setting in which they are located. It presumes that all political actors – voters, professional politicians, bureaucrats – have preferences and make rationally calculated decisions to maximize the realization of their preferences at the least cost. Thus, the different branches of rational choice theory assume that people have preferences and act to attain them.

The study of people to attain their preferences proceeds using four concepts: (1) preferences, (2) strategies, (3) action, and (4) outcomes. Rational political actors develop and execute strategies for action designed to bring about the outcomes they prefer. While the paradigm proceeds from *a priori* assumptions about human behavior, it is capable of generating empirically testable propositions. The specific theories developed out of the rational choice approach thus should be evaluated according to how well they predict or explain behavior, *not* according to how well they correspond to humanistic notions of the complexity of human motivation and social behavior. Like all models or theories, rational choice approaches simplify reality and contain significant built-in biases that need to be reckoned with (see Boyd [1992] and the concluding section of this chapter).

The best developed of the rational choice theories, is economics, the concepts and tools of which have been used to study not only standard behavior in the marketplace but also bureaucracies and even families (McKenzie and Tullock 1975, Putterman 1986). The other branches of rational choice theory of special interest for educational administration are those which deal with modeling social life in terms of 'games,' coalition formation, public choice and voting processes, bargaining, the forecasting of political decisions, and conflict (Riker 1962, Hamburger 1979, Raiffa 1982, Bueno de Mesquita *et al.* 1985, Mueller 1989, Brams 1990, Dixit and Nalebuff 1991).

The richness and diversity of rational choice theory and research are far too extensive to review in depth in this short chapter. The most that can be done here is to give an idea of the field by a quick review of some of its branches and some of the important findings of those branches.

Collective choice theory

The theorists who examine public or collective choice and voting process address a variety of topics, but they all tend to be concerned with 'the action of individuals when they choose to accomplish purposes collectively rather than individually' (Buchanan and Tullock 1962: 13). Collective choice theorists have most centrally looked at the effects of different voting rules on the outcomes of the voting process. What we learn from this work is that *different voting rules – Condorcet, Borda, Bentham – can yield totally different winners when used with the same group of voters holding the same preferences* (Riker 1982). Out of this research came one of the most important theorems in all political science – Black's (1948) median voter theorem – and the discovery of the 'voting cycle' or 'paradox.' To understand Black's theorem, consider the voter whose preference is located at a position where 50% of the voters are located to the left and 50% to the right. The median position is, thus, determined by the point where there are an equal number of voters on each side. It is not necessarily the average position. The theorem holds that under a certain condition (where the voters – e.g., members of a legislature – have single-peaked preference curves) the outcome which coincides with the median voter's position will receive a majority of votes against any other outcome (Strom 1990). The median voter theorem is also used in making predictions regarding the behavior of candidates for elective office – e.g., the best policy position for a politician is the median position (Brams 1985).

The 'so-called voters' paradox or cyclical majority problem, has been discussed by Condorcet, Borda, Dodgson, Black and Arrow since the eighteenth century' (Frohlich and Oppenheimer 1978: 17). The voting cycle is called a paradox because in this case rational voters collectively produce an irrational result, i.e., a set of *intransitive* preferences. Rationality is defined here to mean that if a voter prefers A to B and B to C, then the voter also prefers A to C. These preferences are *transitively* ordered.[2] But suppose three voters – James, Sidney, and Lynn – are faced with three alternatives, A, B, C, upon which they vote in a pairwise comparison, i.e., A is pitted against B, B against C, and A against C. Suppose the preferences of the voters are ranked as shown below.

James	Sidney	Lynn
A	B	C
B	C	A
C	A	B

The preference ranking of these three voters means A gains a majority against B; B wins against C; but C beats A. Collectively, the preference pattern is instransitive or circular; thus any given alternative is defeated by one of the others. In this case it is not possible to know which alternative the voters will select. The existence of this possibility opens the door to agenda manipulations and other sophisticated moves which can be *and are* used by astute politicians to achieve preferred outcomes (Riker 1986, Strom 1990).

Kenneth Arrow (1951) carried our understanding of this paradox one step further when he proved the impossibility of devising a democratic procedure for reaching a group decision that insures transitive and yet nonarbitrary group choices.[3] As Frohlich and Oppenheimer (1978: 27–28) put it, this means that:

> . . . we must face the conclusion that transitivity is sacrificed in democratic governments and examine the implications of that sacrifice. Giving up transitivity leads either to endless cycling and indecision, or to arbitrary choices. Since even democratic governments must make choices, the results must be arbitrary.
>
> The possibility of intransitive social choices means that the order in which issues are put to a vote determines which alternative will be adopted. In any cyclical majority situation (where losers are eliminated), any alternative can get a majority if put to a vote at the appropriate time. Thus, control of the agenda in a cyclical majority situation is tantamount to dictatorial power.
>
> Thus, in an evaluation of a democratic system, we may well focus our attention on the control of agendas, and on whether there are any groups with a monopoly of such control . . . . The relevant questions raised by democracy may not be who votes, but who decides on how the voting is to take place.

The study of the implications of different ways for organizing the agenda for policy outcomes has been central in the research of those investigating collective choice processes (Riker 1986, Strom 1990).

Bargaining and coalition theory

Bargaining theorists have worked on predicting the outcomes of bargaining, the effect of certain bargaining tactics on the (otherwise) predicted outcome, the dynamics of the bargaining process when there are more than two players at the table, the dynamics of mediation and arbitration, and the use of threats (Raiffa 1982, Brams 1990, Young 1991). Coalition theory, which may be seen as a sub-branch of bargaining theory, deals with questions of which coalitions form, what the political goals of that coalition will be, the size of the coalitions, and how the members of the coalition divide the booty they win by becoming a winning coalition. The general approach is to assume that potential coalition partners will join that coalition which leads to the greatest payoff for them; this is when the division of the booty among the members of the winning coalition becomes crucial. One of the principal findings of this body of theory is that if a coalition is big enough to win, it has no need to take on additional members, a finding known as the 'minimum winning coalition' principle (Riker 1962, Young 1991).

Game theory

Researchers addressing both collective choice processes and bargaining concern themselves with political outcomes which depend upon the deliberate strategic choices by two or more actors. This work is thus part of the field called *game theory* which seeks to explain and predict outcomes when players are making their choices in light of the choices other players have made or that they believe the other players will make. When rational choice theorists use the word 'game,' they do not mean games such as football or chess; nor are they referring to simulations or the 'games people play.' What they do refer to are situations in which people find themselves where the 'players' make choices that affect each other. It is from game theory that we get such familiar terms as 'zero-sum,' 'free rider,' 'chicken,' and the 'prisoner's dilemma.'

It may be tempting for the uninitiated to regard game theory as merely an arcane intellectual exercise of doubtful utility. The truth is, however, that game theory, as a way to think about and test political strategies (and ethical rules), has reached even the popular press and government and big business strategizing. For example, in 'A new way to think about rules to live by,' Carl Sagan (1993) discussed game theory in a perceptive and accessible way in *Parade Magazine* (which is not widely known for its intellectual demands). In particular, Sagan drew on Robert Axelrod's (1984) book, *The Evolution of Cooperation*, which shows the superiority of the 'Brazen Rule' over the 'Golden Rule,' etc. For another example, in 'What price air?,' Lewyn (1994) described how not only the FCC, but all the potential bidders (such as Pacific Bell, MCI, Bell Atlantic, etc.) are employing game theorists for strategic guidance in a mega-stakes auction the FCC will hold to allocate licenses for new wireless phone systems known as Personal Communications Services (PCS). Lewyn's article opens by recounting an object lesson, which vividly illustrates why game theorists were being employed:

> Australian bureaucrats thought they had struck it rich last April when they auctioned off rights to use the airwaves for a satellite-TV service. An investment group called Ucom Proprietary Ltd, submitted a sealed bid for $152 million, far above government projections. The problem was that Ucom had no intention of spending that much. It deliberately defaulted on its winning bid, forcing the government to turn to the next lower bid – which was also Ucom's. The company proceeded to default on one bid after another before finally sticking with one that was just slightly higher than the bid of its closest rival, which by then had leaked out. Ucom walked off with the license for $84 million – setting off an uproar that nearly cost Australia's communications minister his job (Lewyn 1994: 48).

Having illustrated the significance of game theory, we can turn to its methods and tools. Game theoreticians have developed a rich vocabulary and a set of graphic devices for describing social phenomena. Sometimes the interaction of players is described using two by two matrices, and sometimes using decision trees with sequences of branches which reflect the dynamic sequence of choices available to one player, then the next, then the first player (Ordeschook 1992, Morrow forthcoming). In addition to their graphic tools, game theoreticians talk in terms of games of complete and incomplete information, games of cooperation and games of conflict, and strategies and equilibrium points. This rich set of concepts and modes of graphic presentation have allowed them to begin to describe a variety of social phenomena in a systematic way.

For example, to describe the emergence of cooperation without coercion, it is useful to introduce a standard way in which game theorists graphically depict the set of choices two individuals may face. The graphic device is a matrix comprising intersecting rows and columns.

| | Player #1 | |
	Column #1	Column #2
Row #1	B, S	T, T
Row #2	T, T	S, B

(Player #2 labels the rows)

One player has the option of choosing between the left- and right-hand columns – choosing a column is a 'move.' The other player chooses between the top row and the bottom row. The intersection of a column and row forms a cell in which there are letters or numbers representing the 'payoff' or outcome for each player. Thus the top left cell contains the payoffs for the two players when the 'column player' has chosen the left column and the 'row-player' has chosen the top row. By convention, the payoffs in the cells are listed with that of the row player first, and the column player's payoff listed after the comma. Also, by convention, the absolute value of the numbers used in the matrix is not significant. What is important is that the relative value of the numbers, i.e., (1, 0) is equivalent to (5, 4), in that the first number is marginally greater than the second number in each set of parentheses. Letters may also be substituted for illustrative purposes, hence B = best payoff; S = second best payoff; T = third best; W = worst payoff.

This matrix format is used to describe a variety of social situations including those in which getting cooperation among the players is a central problem of their relationship. Take the case of two friends who want to do something together on Saturday evening, but Will wants to see a movie and Joe wants to shoot some pool (Luce and Raiffa 1957). Both would prefer to do something together rather than go their separate ways, even if, for example, Will must play pool to be with Joe. This situation is represented in the next matrix.

Reading the matrix, we can see that if Will and Joe agree to go to the movies (Will chooses the top row and Joe chooses the left column), then Will receives a payoff of 2 and Joe a payoff of 1. The situation is different if they agree to play pool – Will receives a benefit of 1 and Joe a benefit of 2. If they go their separate ways they receive no benefit (0, 0). Clearly, the advantage for both is to cooperate, but the issue is cooperation by doing what? The term of the agreement is the point on which they disagree. They could seek to solve the problem by bargaining. One form such an agreement might take is an agreement

	Joe	
	Movies	Pool
Movies	2, 1	0, 0
W i l l Pool	0, 0	1, 2

to go to the movie this Saturday, and play pool the next. Note also that either of the two 'players' could make a *forcing move*. 'I'm going to the movies; join me if you want.' If Will did this, Joe would rationally calculate it would be to his immediate advantage to join Will since he had more to gain by joining Will than by going his separate way (compare Joe's payoff in the left and right columns). Thus, given this situation, we might predict the possibility of a 'forcing move,' and we can predict that they will cooperate in some way, even if the precise terms of the cooperation are not certain without more information about the players and their longer-term relationship.

Notice that this is not a 'zero-sum' game in which one person's gain is another's loss. There are possibilities of mutual advantage in this game (and in another game to be discussed later), as well as conflict of interest. The game is a mixed-motive game in which there is a mixture of conflict and agreement of interests. A significant feature of this situation is that each party can only gain if the other also gains.

There are many situations in the school setting which can aptly be described by this game (sometimes called 'Battle of the Sexes'). School board–union relationships are sometimes aptly described in these terms, as might be the relationship between a supervisor and teacher. Or take the case of teachers engaged in team teaching who need to cooperate on development of the next project the students will undertake. Another more political example is the case of a seven-person school board voting on an issue that requires a two-thirds majority. Suppose four members of the board are willing to vote for a bond issue necessary to raise the money to construct two new buildings, and these board members are indifferent regarding where the larger of the two new buildings will be put. Two more votes are needed to pass the bond issue, but the two board members who are willing to cooperate by voting 'yes' disagree on the location of the buildings. Not to vote 'yes' dooms the project, i.e., everybody goes his or her separate way and the payoffs are zeros. But to vote 'yes' and to pass the bond creates the problem of the terms of the agreement regarding the location of the buildings. One way one of the two board members could force the issue would be to turn-in a proxy vote which specified that his 'yes' vote was contingent upon the larger building being located in a certain place, and then leaving on a vacation where he could not be reached. If the bond issue is to pass, the rationally calculating other yes-vote would have to agree, or suffer the consequence of a zero payoff.

The game just described is just one of many games which have been developed by game theorists which seem to have empirical interest, i.e., they capture real-life interactions that occur with some frequency. Other games that have been described include the 'prisoner's dilemma' game – the game which to date seems to be the most generally applicable game – and the games of 'chicken' and 'convergence.' We shall return briefly to such games in the next section of this chapter.

These games, as noted, can be described using either the matrix format or in the form of a decision tree. Today, for a variety of reasons, game theorists increasingly have opted to use the decision tree to represent their games. The decision tree better represents the sequence of moves and the decision tree's graphic representation lends itself to the addition of many 'bells and whistles' which allow for the effort to represent ever more complex strategic interactions. For example, using the tree, one can easily redesign the tree to change the order of play, i.e., which player moves first; one can represent with certain graphic tools the degree to which the two players know the same thing or must engage in the play with asymmetric information; one can add uncertainty to the game, for example, by allowing 'nature' to make the first move and not letting any of the players be sure exactly what the move is that nature made. In short, the decision tree has become a very sophisticated vocabulary that allows the construction of models that take into account an ever more complex set of social situations (Morrow forthcoming).

The advantages to the political scientist of trying to describe reality using the vocabulary of these models are multiple. It provides a way of being very systematic in relating the relevant variables one is concerned with to each other – one can see and understand fully the effect of independent variable X on the outcome. The researcher can systematically think through the likely logical effects of a change in one variable, e.g., change the game so that Jane moves before Joe. Thus one can develop hypotheses regarding the effects of changes of variables. These models are typically further specified in the language of mathematics, and the logic of that discipline helps the researcher work systematically through the effects of changes in variables with great precision. Once developed the model, to an extent, takes on a life of its own and the research is led to new insights and possibilities regarding 'reality' which then can be confirmed or discomfirmed through empirical research. The model can act as a kind of flashlight illuminating features of the situation one had not noticed when looking at the actual data. Having suggested that something may be found in the data, the modeler can now return to the data set to see if what the model suggested is in fact empirically born out. For example, if you look back at the 'battle of the sexes,' the basic matrix suggested the possibility of a *forcing move* by one of the players. The modeler might not have realized this possibility before drawing the matrix. Having constructed the matrix the forcing move becomes an obvious possibility. Thus the modeler can return to the data set, and see if a forcing move occurred. Now a series of additional questions present themselves – when is or is not the forcing move a real possibility? What if we assume the two have an ongoing relationship as opposed to a single-event relationship? How does that affect our prediction that a forcing move will occur? Thus, in this way the modeler explores and generates new hypotheses for further testing.

Review of research using the rational choice paradigm

As noted earlier, little research on educational politics has been conducted using rational choice, and none of this research to date has actually relied on formal model building and testing. However, some research exists that can be seen to have been influenced by rational choice theory, some of which focuses upon educational issues, and some of which deals with other parts of the political system but is nevertheless of direct relevance to understanding educational politics.

Collective choice and legislative processes

No topic has been examined more closely by rational choice theorists than collective choice processes and, especially, legislative processes. These investigations have focused on the effects of different voting and procedural rules upon the behavior of voters, legislators, and legislative outcomes. In fact, one of the significant findings of this body of work is that procedural rules commonly found in legislative bodies, such as the division of the question rule (under which a complex piece of legislation involving, say, two distinct issues, is divided into two parts with a separate vote held on each part), make it possible for there to be a stable legislative outcome reflective of the preferences of the median voter-position on each issue. In the absence of such procedural rules, there might not be a stable outcome, as the legislature might get caught in the voting cycle discussed earlier (Strom 1990).

Two important phenomena of the legislative process discovered and analyzed by rational choice theorists are the 'saving' and 'killer' amendments. Any amendment to an original bill has the effect of placing before the legislature three options from which to choose: the original bill, the amended bill, and the status quo. Depending upon the distribution of preferences of the legislators, these three options create the possibility of a 'voting cycle' that can be exploited. If such a cycle is possible, astute legislators can use a 'saving' amendment to exploit an intransitive voting cycle to save a bill that otherwise would have gone down to defeat. A 'killer' amendment creates the same voting cycle, but is exploited to lead to the defeat of the original bill (which otherwise would have passed) and the maintenance of the status quo.

Professor William Riker (1986: 114–128) described the operation of a killer amendment in conjunction with a school aid bill before Congress in 1956. The original bill would have provided federal aid for the construction of school buildings, but Representative Adam Clayton Powell, an African-American elected from New York City, introduced an amendment to the bill that required that money be given only to schools which did not discriminate on the basis of race. Powell introduced the amendment to force members of the House to go on record in favour of racial integration. The effect of the amendment, however, was to create a voting cycle involving the original bill, the amended bill, and the status quo (no federal aid for construction). The Republicans in the House, who opposed the original bill, saw their opportunity. Although they preferred the original bill to the amended bill, in the first vote, which put the original bill against the amended bill, they voted 'insincerely' for the amended bill which passed. Thus, the final vote was between the amended bill and the status quo. Now the Republicans joined the Southern Democrats to form a majority which defeated the amended bill; thus the status quo prevailed. Powell may have achieved his symbolic victory, but his amendment operated as a 'killer' amendment.

Political participation and the 'free rider' problem

One virtue of the rational choice approach is that it calls attention to a number of important costs (e.g., information, opportunity, and transaction costs) involved in political behavior that the traditional approaches of political scientists neglected or underestimated. These costs are usefully identified in a number of related concepts that can be applied, for example, to the study of voting behavior (Downs 1957), and to the problems of mobilizing political action and forming and maintaining interest groups (Olson 1965). Such notions can be related in propositions that illuminate the difficulties of citizen participation in education policy making (Weeres 1971, Peterson 1974). For instance, the costs of obtaining information, of participating in decision making, and of mobilizing political action all increase as the size of a governmental unit, such as a school district, increases (Black 1974). Similarly, the costs of an individual's foregone opportunities, and the small probability that one's own actions, by themselves, will significantly affect the ability of a large organization to achieve some general objective, tend to discourage individuals from contributing to the organization's efforts. Moreover, if the objective should be achieved, it is sometimes the case that nonmembers of the organization will benefit as much as members.

Such is clearly the case with public or collective goods, such as public education, and this raises what economists call the 'free rider' problem (Olson 1965). A public or collective good is one that, if provided, is available to everyone and no one can be excluded. Clean air and national defense are examples of such goods. The difficulty with this property of collective goods is that it creates a situation where it is not to the advantage of any rational, self-interested person to contribute to the provision of the good. Instead, it is to his or her advantage to take a free ride, enjoying all the benefits of the collective good without incurring any of the costs of providing it. The result, of course, is that collective goods ordinarily will be provided only if government undertakes them and *compels* everyone to contribute to their costs. This principle also applies to private or voluntary organizations, which must be able to compel or specially reward contributions to their efforts. The insights into the problems of collective action and, indeed, the likelihood of collective *inaction* under certain circumstances revealed by this line of analysis forced a major reassessment and reformulation of the body of interest group theory that for so long guided much of American political science (see, e.g., Wilson 1973).

Goals, rules and self-interest

Another virtue of the rational choice approach is that the paradigm immediately calls attention to the difference – and tension – between the goals of individuals (maximizing their own welfare) and the professed goals of organizations. As a consequence, it provides a rational explanation for much behavior that otherwise appears irrational or pathological in terms of the announced goals of organizations. In a profit-seeking, private-sector organization facing competition, the incentive structure of the organization motivates self-interested employees to engage in behaviors convergent with the maximization of profits. This structure necessitates satisfying customers, thereby presumably achieving (or at any rate approaching) the organization's announced goals. In the case of public schools, the primary announced goal is the production of valued student learning outcomes. But, as Jacob Michaelsen (1977) and others have pointed out, public schools are quasi-monopolistic, nonprofit government agencies whose financial support comes through a political process in the form of a tax-supplied budget rather than directly from satisfied clients. Thus, the crucial linkage that insures consumer sovereignty is broken, because of the assurance of a budget *independent* of the degree to which individual consumers are satisfied. Since there are no profits in public agencies (including public schools) to motivate and reward managers (and teachers' salaries are based on seniority rather than performance), Michaelson (1977: 329) contends that, in place of profits,

> . . . we may assume that bureaucrats, including schoolmen, seek instead to survive, to enlarge the scope of their activities, to gain prestige, to avoid conflict, to control the organization and content of their daily round as much as possible. All these are, as it were, profit in kind. By accepting this theory of motivation, we do not rule out

altruism. Administrators and teachers, like entrepreneurs, take pride in their work and strive for excellence. The issue is not whether bureaucrats are altruistic but rather whether there are mechanisms available to harness their self-seeking to the public interest.

In the contest between individual goals and the announced goals of school systems, Michaelsen (1977) and others have shown how public employees stand to gain greater benefits and face lower costs (e.g., lower information, opportunity, and political mobilization costs) in pursuing their personal goals ('unofficial benefits') from the organization than the lay public faces in seeking to insure the attainment of the announced goals (or 'official benefits') of the school system. Downs (1957) and Niskanen (1971) also have shown the vulnerability of legislatures (e.g., school boards) to manipulation and lobbying by bureaucratic executives and employees. Furthermore, the exclusivity of unofficial benefits and the non exclusivity of official benefits create substantial difficulties (especially the 'free rider' problem) in mobilizing collective action (i.e., in forming and maintaining interest groups) for the public that are not faced by employees or others interested in the unofficial benefits. Taken together, the rational choice paradigm provides a comprehensive explanation for commonly observed differences between *consumer* groups (the lay public) and *producer* groups (i.e., public employees, such as public educators) in political organization and action in behalf of their respective goals.

Interestingly, Chubb and Moe's (1990) controversial analysis took the next logical step beyond the critique of 'producer' capture of the public schools: They concluded that political governance of schools is unlikely to work effectively, *regardless* of whether the lay public or the producer groups control it. Their evidence, they contend, suggests that school effectiveness depends upon schools being autonomous, free of the overburden of political control and excessive regulation they believe typifies American public schools. The heart of the problem, they assert, is that political control ensures a steady growth of burdensome bureaucratic regulations, as interest groups (and not least the 'producer' groups) use the process to protect and advance their policy preferences. Thus, in a startling conclusion for political scientists, they call for an end to democratic control of schools. (This is a growing trend: see Plank and Boyd 1994.) Rather, they say, the public schools should be placed squarely into a market system. Recognizing the likely inegalitarian consequences of an unregulated 'free market,' they advocate a 'regulated' market for schools, which places them on a slippery slope from the perspective of free market economists: How much regulation would be enough, and how much too much, either in economic terms or in terms of their own theory of school autonomy?

Game theory

Whether trying to design reforms, improve schools or simply choose political tactics or strategies that will succeed, reformers and practitioners often face situations in which the outcomes depend, not simply on their choices, but 'on the deliberate choices of other rational decisionmakers' (Little 1991: 51–52):

> This is a situation of *strategic* rationality . . . [in which] the payoff to the individual depends on the choices made by the other players. So each decisionmaker must consider the rational calculations of the others and choose that option that maximizes his or her payoff *given* the assumption that all the others make a rational decision as well. (Little 1991: 52)

Game theory illuminates this arena and its special problems, such as those embodied in the famous 'prisoner dilemma' situation, in which two thieves who are accomplices are interrogated separately and end up incriminating each other, out of fear the other will 'talk,' even though both would have been better off remaining silent. Von Neumann and Morgenstern (1944) developed game theory to better understand their favorite game – poker. Given a particular hand, their goal was to determine the best strategy to pursue – bluff, fold, stand pat, or raise. Tyll van Geel (forthcoming), who now offers a course for educational leaders based on rational choice theory, explains the applicability to education of various game situations, such as the famous 'chicken' game:

> The game is best illustrated by two automobiles driving at great speed in opposite directions directly at each other. The first driver to swerve is the 'chicken,' but failure of either to swerve spells disaster for both . . . . The game is a useful way to illustrate difficult confrontations, as occurred between the school board of the City School District of Rochester, New York, and the teachers union headed by Adam Urbanski. The confrontation arose over the need for the board to make a mid-year reduction in the school budget. The board favored cutting salaries and services at the school building level, e.g. elementary school librarians, and a give-back of teacher salaries, while the union favored reductions in central office staff. Failure of either party to swerve would send the board, union and community into further debt, the need to borrow money and a down-grading of the district's credit rating. Yet for each party the rational choice may be to be hard-nosed. School board members faced legal and other pressures that made it impractical to engage in a full scale elimination of central office jobs. Union officials faced loss of position in the next union election. But, certainly, the pressure to reach an agreement was severe, for failure to do so would have landed the two players (as well as the community) in the [disaster] cell. Negotiations dragged on for some months until finally the board was forced to act unilaterally imposing a variety of steps that led to the laying off of approximately 150 people. The board's relationship with the union remains difficult a year after these events. (van Geel forthcoming: 12)

Organizational economics

The relatively new field of organizational economics (Moe 1984, Barney and Ouchi 1986) is full of insights for those who desire to reform or better understand educational organizations. Indeed, it is one of several areas that deserve much more attention than we can give them here. Interestingly, at this time of de-bureaucratization, the organizational economics literature directs attention anew to questions of organizational hierarchy. The key construct is the 'principal-agent' relationship – wherein it is assumed that agents up and down an authority structure are 'induced' to pursue the objects and expectations of the superordinates (i.e., their 'principals') (Moe 1984, Stiglitz 1987).

Agency theory offers an initial insight of value into just why individuals join organizations in the first place (the employment 'contract'). The argument is that by entering into authority relations, usually arranged hierarchically, individuals save themselves the 'costs' of coordinating their own productivity (Coase 1937). Classroom teachers, for example, as autonomous as they are or would wish to be, depend heavily upon the personal 'cost-savings' provided by an organizational structure that identifies and enrolls pupils, provides and maintains facilities, allocates space, distributes teaching resources, guards employee and client safety, and imposes a common calendar.

A second contribution of importance from agency theory is its full recognition of, and rational explanation for, the 'messiness' of the employment contract – of the deep conflicts of interest, the 'asymmetries' in the information available to principals and their agents, the many ambiguities regarding rewards and performance expectations, and the added costs of effort to monitor the 'compliance' of agents (Moe 1984). In 'economics of organization' terms, goal displacement or an under-supply of effort can be highly rational behaviors – as McCubbins et al. (1987) illustrate in noting that public-sector boards of control tend to be reactive in monitoring their administrative agents (attending especially to 'fire alarms'). Thus, the chief administrators – never certain as to just when and for what reasons an alarm might go off – tend to be extremely cautious and conservative.

Finally, agency theory represents a uniquely open and political approach to the economics of organizational hierarchy – in that it extends to its notions of principal–agent relationship in 'systemic' fashion, far 'above' and 'below' the narrowly institutional. The point is nicely illustrated for education in Jane Hannaway's (1988) discovery that district-level school administrators rated tasks they perceived to be of state and federal origin as 'more important' than work tied to their own districts' key 'production' concerns (i.e., teaching and learning). Similarly, Galvin (1993, forthcoming) uses organizational economics perspectives to illuminate cross-boundary decision making about cooperation among educational organizations, and shows the naiveté of arguing that cooperation among such organizations is inevitably a good thing. Since cooperation can be hard to achieve and can have a political dimension (Blase 1991), the insights of organizational economics can benefit studies in the politics of education.

Political forecasting

Although not yet applied to educational politics, the new tools of political forecasting have the potential to make valuable contributions. Some of the most interesting and advanced work on political forecasting and conflict is being done today by Bruce Bueno de Mesquita, based on an 'expected utility' approach to understanding human choices. Based on his mathematical models and using a minimum of political data, Bueno de Mesquita can and has predicted the outcomes of such political decision-making processes as the negotiations between Great Britain and the Republic of China over the fate of Hong Kong, and the negotiations among the members of the European Union over such matters as the year in which they agree that certain pollution control requirements will go into effect on automobiles (Buena de Mesquita et al. 1985, Bueno de Mesquita forthcoming). Bueno de Mesquita has developed related models which he uses to forecast the outcomes of international conflicts, e.g., whether two nations will go to war, bargain, and even whether one or the other side will yield or capitulate to the other (Buena de Mesquita and Lalman 1992). These models can be used in connection with domestic conflict and domestic political decision-making, even at the level of the local school board.

Policy implications of rational choice research

The 'market model' and 'rational choice' critiques of the monopolistic pathologies of public sector organizations and, indeed, of the welfare state itself have had profound consequences for public policy and politics, not just in the USA but worldwide (Boyd 1992). Acceptance of these critiques, and their growing influence, were accelerated by the worldwide economic problems beginning with the OPEC oil embargo and associated 'stagflation.' In the public education sector, along with mounting concern about declining student performance, and about the deterioration of the moral order of schools, a substantial part of the continuing 'school choice' or 'voucher plan' debate has been fueled by these critiques. More broadly, economic models and related productivity concerns have substantively influenced the policy debate on American school reform.

For instance, whatever one may think of Chubb and Moe's (1990) analysis, or of voucher plans, there is no denying that the newest growth industry in policy design for American education involves efforts to free schools from excessive bureaucratic and political control, while at the same time creating new and, hopefully, effective accountability mechanisms to harness educators' (and students') efforts. The rapid growth of the 'charter schools' movement, the new and related 'contract schools' movement championed by Paul Hill (1994), the privatization and 'contracting out' movement epitomized by Educational Alternatives, Inc. and by Whittle Communication's Edison Project, and diverse efforts to reinvent American schools and, above all, to revive or replace our failing large-city school systems all exemplify the new growth industry.

The accountability and outcomes or results-oriented strand of these developments owes a good deal to the economic/rational choice critique of schools as they formerly and, sadly, too often still exist: Where student outcomes are viewed as mainly, or sometimes entirely, the responsibility of children and their families, there is no need to 'grade' schools or teachers. From an economic or rational choice perspective, the resulting low cost to educators for inefficient behavior creates, in effect, a 'demand for inefficiency,' permitting educators not otherwise driven by altruism to wallow in unaccountable sloth (see Chambers 1975, McKenzie and Tullock 1975). At the same time, with 'social promotion' students can behave 'rationally' by making minimal efforts at learning since they know they can graduate from high school largely by just putting in enough 'seat time.' Unless these incentive problems for students and educators are corrected, economic analysis suggests that efforts to improve school performance may amount to little more than 'throwing money at schools' (Hanushek 1981, 1986, see also Boyd and Hartman 1988).

The prospects and limits of rational choice theory

The rational choice literature has established a number of theorems and concepts of great power – e.g., the median voter theorem, the Coase theorem, Bank's monotonicity theorem, the Nash concept of equilibrium, the Bayesian theorem, expected utility theory, and, as noted earlier, Arrow's theorem. These theorems, concepts, and theories are today being creatively combined to develop other powerful theories of human behavior which have been empirically tested and which yield remarkably accurate predictions. One of the best of these new models is the model, noted earlier, of political forecasting developed by Bueno de Mesquita. Strides have also been made in using rational choice models to understand the operation of hierarchical organizations (Bacharach and Lawler 1980, Moe 1984, Miller 1992), and the impact of law (Posner 1992).

Despite the enormous progress that rational choice theory has made, it is important to remember that this theory is still in its infancy and, thus, still has important limitations. Some of the models do not yet yield the kind of finite predictions of the Bueno de Mesquita model, i.e., the models point to a variety of possible outcomes, but do not specify which will in fact be the outcome. Thus, in some cases it is still possible to use rational choice thinking to arrive at seemingly contradictory conclusions, e.g., that democracy in the workplace will lead, on the one hand, to instability and inefficiency, but, on the other hand, to better, more efficient decisions (Miller 1992: 64, 81). It is also true that people do not always behave the way some rational choice models say they will. Take, for example, bargaining: rational choice models of bargaining tend to predict that the bargainers will reach efficient decisions and will extract the most each can out of the deal. But these predictions are confounded by social psychological experiments and observation of real life which show people acting 'irrationally' and making mistakes in bargaining (Young 1991). It is also true that this mode of analysis has not yet been extended to explain a variety of social phenomena. These models do not seek to explain why people have their basic preferences – the models take these preferences as a 'given.' There are also phenomena in social life that these models have had difficulty explaining. Curiously, one of the more difficult questions of social life, which the theory has not yet fully answered, is why people vote. Why does a rational person take the time, energy, and run the possible (if unlikely) risks of an accident on the way to the poll when his or her vote has so little chance of having a meaningful impact among the millions of other votes cast?

As rational choice modelers address questions such as these, they confront the issue of whether the world should always be looked at in individualistic terms. In other words, is it the case that 'Rational choice is about how individuals make choices, and sociology is about how individuals have no choices to make' (Miller 1992: 206)? While the difference between the disciplines is narrowing, as sociologists take up rational choice (e.g., Coleman 1990), and as rational choice theorists become concerned with, for example, 'corporate culture,' and 'conventions,' many critics remain who argue that rational choice models are severely limited because, in their view, they are simplistic, seriously incomplete, or inappropriate or wrong-headed for educational institutions (e.g., the emphasis on competition despite much evidence that successful schools stress cooperative relations among students and teachers).

Without doubt, economic approaches do contain significant biases that need to be reckoned with (Boyd 1992). The economic paradigm, and the 'technocratic' policy analysis often associated with it, are certainly not value-free. Indeed, Laurence Tribe's warning in 1972 about the ideological dangers embedded in these approaches remains valid:

> [T]he policy sciences' intellectual and social heritage in the classical economics of unfettered contract, consumer sovereignty, and perfect markets both brings them within a paradigm of conscious choice guided by values and inclines them, within that paradigm, toward the exaltation of utilitarian and self-interested individualism, efficiency, and maximized production as against distributive ends, procedural and historical principles, and the values . . . associated with personal rights, public goods, and communitarian and ecological goals. (Tribe 1972: 105)

Tribe's fears were borne out by the trend of American social affairs. Following on the heels of disillusionment with the Great Society and War on Poverty programs of the 1960s, the economic problems of the 1970s brought a resurgence of conservatism in America and an associated rehabilitation of the free market metaphor. In the reaction against the perceived failure of government social programs, the 'Great Society' envisioned by liberal sociologists was replaced by what can be called the 'Fragmented Society,' a society shaped by the views of economists who stressed the need to design policy to harness the energy of 'self-interest maximizing individuals' (Schultze 1977, Bellah 1983).

In this context, policy analysis guided by the economic paradigm of the 'self-interest maximizing individual' itself contributed to the dramatic shift in the nature and semantics of American discourse about social policy and the public interest during the Reagan and Bush years. Dialogue and concern moved from equity, social justice, and the common good to questions of liberty, choice, excellence, and efficiency. In education, this trend was particularly evident in the erosion of support for common, public schools and concomitant interest in private schools, tuition tax credits, and voucher plans. All of these developments reflected a decline of concern for community and the legitimacy that the self-interest paradigm gave to competitive struggle. What happened was just what the proponents of value-critical policy analysis (e.g., Rein 1983, Prunty 1984) warned against: The pervasive paradigm of pursuit of self-interest eroded and deflected attention from the value of *community*.

By the time Bill Clinton was elected president, there were signs that a synthesis might be emerging between the extremes of the 'great society' and the 'fragmented society.' The 'communitarian' movement had formed in response to the erosion of a sense of commonweal (Etzione 1988, 1993). Policy analysts and reformers wishing to 'reinvent' government were seeking ways to avoid the twin dangers of market failure and government failure (Weimer and Vining 1989, Osborne and Gaebler 1992). Diatribes against public school monopolies were being replaced by more practical policy proposals, such as charter schools and contract schools. The debate, at least among mainstream groups, had moved to accept part of the rational choice critique of government schools and the welfare state, but tempered it with revived concerns about community, equity, and social justice.

With respect to work done in the rational choice tradition, two main kinds of challenges have been levelled. One set of challenges is really aimed, as noted above, at economics-based policy analysis. Thus, the use of cost–benefit analysis has been challenged because it ignores matters such as inequalities in the distribution of wealth, and the prescriptions which arise from cost–benefit analysis may aggravate inequalities in wealth distribution. The other kind of challenges are the sorts of issues taken up in Daniel Little's (1991) superb book – challenges to the philosophy of science behind the kind of work, for example, that political forecasters such as Bueno de Mesquita do. Thus, these challenges are more along the line that this sort of science is culturally biased and only useful for explaining phenomena in certain Western cultures and, at that, not all phenomena in western culture. One answer to that charge is proof that the approach does explain and predict better than the critics charge.

A different kind of charge that critics make is that rational choice does not really capture 'the meaning' of events from the subjective perspective of the participants. But the true test of rational choice models is whether they offer sound explanations of phenomena employing a real causal mechanism, and whether they are predictive of a wide range of phenomena. If the models do not explain and/or are not predictive, then they need to be improved. But since the kind of work that many rational choice theorists are trying to do is descriptive and predictive, the criteria by which it should be assessed are not ideological, but those which arise out of the canons of science and philosophy of normal science. Of course, one can challenge the canons of the philosophy of behavioral science, but then one spins off into issues of epistemology and the kinds of problems taken up by Little (1991).

Conclusion

The limited use of rational choice theory by scholars in education schools derives, in part, from characteristics of schools of education and, in part, from perceived limitations of this body of theory. Very few faculty members in schools of education have been trained in or exposed to economics. It is not uncommon even for people teaching school finance or business management to have had little economic training. The low visibility of economics in education schools is not entirely accidental: The perspectives of business, economics, and administration tend to diverge from, and not easily harmonize with, the professional ideology of teachers and their mentors. For example, tensions between teachers and school administrators are sometimes echoed in schools of education. Moreover, the ideology of education schools, and the self-interest of their inhabitants, make 'improvement' of schools acceptable, but not 'criticism' of schools. ('Don't bite the hand that feeds you'.) In this context, the economic critique of public schools as prone to monopolistic pathologies is nearly treasonable. The accepted view is that public educators are – or at any rate *ought* to be – almost entirely driven by altruism. Hence, it follows that the monopolistic, self-interest critique must be wrong.

Still, the mounting calls for 'systemic reform' and the 'reinvention' of American education are being heard even inside schools of education. So, even if they disagree with the rational choice critique of public schools as institutions, education school faculty are increasingly open to serious discussions about fundamental restructuring of schools. These discussions are most likely to take a sociological approach, emphasizing the need for enhancing and professionalizing the workplace of teachers. Inevitably, however, these discussions will need to contend with the issue of incentives, which opens up the possibility of analysis and action more in accord with rational choice theory.

Thus, at this juncture, the most likely predictions about the future role and impact of rational choice theory in the study of education politics is that it will gradually gain more attention and use simply because of its growing acceptance and use in the other social sciences, not to mention its rather central role in policy analysis. It waits in the wings for those venturesome enough to exploit its potential in the field of education.

Notes

1. For other examples of the use of rational choice theory in the analysis of education politics, see Boyd (1980, 1982), Boyd and Crowson (1981), Michaelsen (1981), Pincus (1974), Shapiro and Crowson (1986), van Geel (1978), West (1967).
2. One of the cases Riker documents (described later in this chapter) involved a 1956 bill for federal aid to education.
3. Transitivity exists when all alternatives can be ranked unambiguously from most to least preferred. Sports fans are familiar with the problems when transitivity is lacking: 'If we try to rank three football teams when Texas beat Arkansas, Arkansas beat Oklahoma, and Oklahoma beat Texas, we will have trouble deciding which is to be ranked first' (Frohlich and Oppenheimer 1978: 7).
4. It was principally for this accomplishment that Arrow received the Nobel Prize.

References

ALCHIAN, A. and ALLEN, W. R. (1977) *Exchange and Production: Competition, Coordination, and Control*, 2nd edn (Belmont, CA: Wadsworth).

ARROW, K. J. (1951) *Social Choice and Individual Values* (New Haven, CT: Yale University Press).

AXELROD, R. (1984) *The Evolution of Cooperation* (New York: Basic Books).

BACHARACH, S. B. and LAWLER, E. J. (1980) *Power and Politics in Organizations* (San Francisco: Jossey-Bass).

BARNEY, J. B. and OUCHI, W. G. (eds) (1986) *Organizational Economics* (San Fransicso: Jossey-Bass).

BELLAH, R. N. (1983) Social science as practical reason, in D. Callahan and B. Jennings (eds) *Ethics, the Social Sciences, and Policy Analysis* (New York: Plenum Press).

BLACK, D. (1948, February) On the rationale of group decision-making, *Journal of Political Economy*, 56: 23–34.

BLACK, D. (1958) *The Theory of Committees and Elections* (Cambridge: Cambridge University Press).

BLACK, G. (1974, September) Conflict in the community: a theory of the effects of community size, *American Political Science Review*, 68(3): 1245–1261.

BLASE, J. (ed.) (1991) *The Politics of Life in Schools: Power, Conflict, and Cooperation* (Newbury Park, CA: Sage).

BOYD, W. L. (1980) The political economy of education in metropolitan areas: dilemmas of reform and public choise, *Educational Evaluation and Policy Analysis*, 2(1): 53–60.

BOYD, W. L. (1982) The political economy of public schools, *Educational Administration Quarterly*, 18(3): 111–130.

BOYD, W. L. (1992, November) The power of paradigms: reconceptualizing educational policy and management, *Educational Administration Quarterly*, 28(4): 504–528.

BOYD, W. L. and CROWSON, R. L. (1981) The changing conception and practice of public school administration, in D. Berliner (ed.), *Review of Research in Education*, Vol. 9 (Washington, DC: American Educational Research Association), 311–373.

BOYD, W. L. and HARTMAN, W. (1988) The politics of educational productivity, in D. Monk and J. Underwood (eds) *Microlevel School Finance: Issues and Implications for Policy* (Cambridge, MA: Ballinger).

BRAMS, S. J. (1985) *Rational Politics* (Washington, DC: Congressional Quarterly).

BRAMS, S. J. (1990) *Negotiations Games: Applying Game Theory to Bargaining and Arbitration* (New York: Routledge).

BUCHANAN, J. M. and TULLOCK, G. (1962) *The Calculus of Consent: Logical Foundations of Constitutional Democracy* (Ann Arbor, MI: University of Michigan Press).

BUENO DE MESQUITA, B. (forthcoming) Political forecasting: an expected utility method, in B. Bueno de Mesquita and F. Stokman (eds) *Twelve into One: Models of Decision-making in the European Community* (New Haven, CT: Yale University Press).

BUENO DE MESQUITA, B. and LALMAN, D. (1992) *War and Reason* (New Haven: Yale University Press).

BUENO DE MESQUITA, B., NEWMAN, D. and RABUSHKA, A. (1985) *Political Forecasting: The Case of Hong Kong* (New Haven: Yale University Press).

CHAMBERS, J. G. (1975) An economic analysis of decision-making in public school districts, unpublished paper, Graduate School of Education and Human Development, University of Rochester.

CHUBB, J. E. and MOE, T. M. (1990) *Politics, Markets and American Schools* (Washington, DC: Brookings Institution).

COASE, R. H. (1937) The nature of the firm, *Economica*, 4: 386–405.

COLEMAN, J. S. (1990) *Foundations of Social Theory* (Cambridge, MA: Harvard University Press).

DIXIT, A. and NALEBUFF, B. (1991) *Thinking Strategically* (New York: W. W. Norton).

DOWNS, A. (1957) *An Economic Theory of Democracy* (New York: Harper & Row).

ETZIONI, A. (1988) *The Moral Dimension: Toward a New Economics* (New York: Free Press).

ETZIONI, A. (1993) *The Spirit of Community: Rights, Responsibilities, and the Communitarian Agenda* (New York: Crown).

FROHLICH, N. and OPPENHEIMER, J. A. (1978) *Modern Political Economy* (Englewood Cliffs, NJ: Prentice-Hall).

GALVIN, P. F. (1993) Black boxes, contingencies and quasi-markets: a theoretical analysis of cooperation among educational organizations, in *Advances in Research and Theories of School Management and Educational Policy*, Vol. 2 (Greenwich, CT: JAI Press), 25–60.

GALVIN, P. F. (forthcoming) Cooperation among school districts: issues of effectiveness and responsibility, chapter in book edited by R. Stephens.

HAMBURGER, H. (1979) *Games as Models of Social Phenomena* (San Francisco: W. H. Freeman).

HANNAWAY, J. (1988) *Managers Managing: The Workings of an Administrative System* (London: Oxford University Press).

HANUSHEK, E. A. (1981) Throwing money at schools, *Journal of Policy Analysis and Management*, 1(1): 19–42.

HANUSHEK, E. A. (1986) The economics of schooling: production and efficiency in public schools, *Journal of Economic Literature*, 24(3): 1141–1177.

HILL, P. (1994) Reinventing public education. *Phi Delta Kappan* (January): 396–401.

LASSWELL, H. D. (1936) *Politics: Who Gets What, When, How* (New York: McGraw-Hill).

LEWYN, M. (1994, March 14) What price air?: the FCC readies an airwave auction by boning up on game theory, *Business Week*, 48: 53–54.

LITTLE, D. (1991) *Varieties of Social Explanation: An Introduction to the Philosophy of Social Science* (Boulder: Westview Press).

LOWI, T. J. (1992, March) The state in political science: how we become what we study, *American Political Science Review*, 86: 1, 1–7.

LUCE, R. D. and RAIFFA, H. (1957) *Games and Decisions* (New York: Wiley).

McCUBBINS, M. D., NOLL, R. G. and WEINGAST, B. R. (1987) Administrative procedures as instruments of political control, Working Paper 109, Center for the Study of American Business, Washington University, St Louis, MO.

McKENZIE, R. B. and TULLOCK, G. (1975) *The New World of Economics* (Homewood, IL: Richard D. Irwin).

MICHAELSEN, J. B. (1977) Revision, bureaucracy, and school reform, *School Reviews*, 85 (February): 229–246.

MICHAELSEN, J. B. (1981) A theory of decision-making in the public schools: a public choice approach, in S. B. Bacharach, (ed.), *Organizational Behavior in Schools and School Districts* (New York: Praeger), 208–241.

MILLER, J. G. (1992) *Managerial Dilemmas* (Cambridge: Cambridge University Press).

MOE, T. M. (1984) The new economics of organization, *American Journal of Political Science*, 28(4): 739–777.

MORROW, J. D. (forthcoming) *Game Theory for Political Scientists* (Princeton, NJ: Princeton University Press).

MUELLER, D. C. (1989) *Public Choice II* (Cambridge: Cambridge University Press).

NISKANEN, W. A. (1971) *Bureaucracy and Representative Government* (Chicago: Aldine).

OLSON, M., Jr (1965) *The Logic of Collective Action: Public Goods and the Theory of Groups* (Cambridge, MA: Harvard University Press).

ORDESCHOOK, P. C. (1992) *A Political Primer* (New York: Routledge).

OSBORNE, D. and GAEBLER, T. (1992) *Reinventing Government: How the Entrepreneurial Spirit is Transforming the Public Sector* (Reading, MA: Addison-Wesley).

PETERSON, P. E. (1974) Community representation and the 'free rider', *Administrator's Notebook*, 22(8): 1–4.

PETERSON, P. E. (1976) *School Politics, Chicago Style* (Chicago: University of Chicago Press).

PETERSON, P. E. (1981) *City Limits* (Chicago: University of Chicago Press).

PINCUS, J. (1974) Incentives for innovatin in the public schools, *Review of Educational Research*, 44, 113–144.

PLANK, D. and BOYD, W. L. (1994, Summer) Anti-politics, education, and institutional choice: the flight from democracy, *American Educational Research Journal*, 31(2): 263–281.

POSNER, R. (1992) *Economic Analysis of the Law*, 4th edn (Boston: Little, Brown).

PRUNTY, J. J. (1984) *A Critical Reformation of Educational Policy and Policy Analysis* (Geelong, Victoria, Australia: Deakin University Press).

PUTTERMAN, L. (ed.) (1986) *The Economic Nature of the Firm: A Reader* (Cambridge: Cambridge University Press).

RAIFFA, H. (1982) *The Art and Science of Negotiation* (Cambridge, MA: Belknap Press of Harvard University Press).

REIN, M. (1983) Value-critical policy analysis, in D. Callahan and B. Jennings (eds) *Ethics, the Social Sciences, and Policy Analysis* (New York: Plenum).

RIKER, W. H. (1962) *The Theory of Political Coalitions* (New Haven, CT: Yale University Press).

RIKER, W. H. (1982) *Liberalism Against Populism* (San Francisco: W. H. Freeman).

RIKER, W. H. (1986) *The Art of Political Manipulation* (New Haven: Yale University Press).

SAGAN, C. (1993, November 28) A new way to think about rules to live by, *Parade Magazine*, 12–14.

SCHULTZE, C. L. (1977) *The Public Use of Private Interest* (Washington DC: Brookings Institution).

SHAPIRO, J. Z., and CROWSON, R. L. (1986) Rational choice theory and administrative decision-making: implications for research in educational administration, in S. B. Bacharach (ed.), *Advances in Research and Theories of School Management* (Greenwich, CT: JAI Press), 279–302.

STIGLITZ, J. E. (1987) Principal and agent, in J. Eatwell, M. Milgate and P. Nennan (eds) *The New Palgrave Dictionary of Economics* (London: Macmillan).

STROM, G. S. (1990) *The Logic of Lawmaking* (Baltimore: Johns Hopkins University Press).

TRIBE, L. H. (1972) Policy science: analysis or ideology?, *Philosophy & Public Affairs*, 2(1): 55–110.

VAN GEEL, (1978) Parental preferences and the politics of spending public educational funds, *Teachers College Record*, 79, 339–363.

VAN GEEL, T. (forthcoming) The preparation of educational leaders and rational choice theory, in R. Donmoyer, M. Imber and J. Scheurich (eds) *The Knowledge Base in Educational Administration: Alternative Perspectives* (Albany, NY: SUNY Press).

VON NEUMANN, J. and MORGENSTERN, O. (1944) *Theory of Games and Economic Behavior* (Princeton, NJ: Princeton University Press).

WEERES, J. G. (1971) School politics in thirty-three of the local community areas within the City of Chicago, unpublished PhD dissertation, University of Chicago.

WEIMER, D. L. and VINING, A. R. (1989) *Policy Analysis: Concepts and Practice* (Englewood Cliffs, NJ: Prentice-Hall).

WEST, E. G. (1967) The political economy of American public school legislation, *Journal of Law and Economics*, 10, 101–128.

WILSON, J. Q. (1973) *Political Organizations* (New York: Basic Books).

YOUNG, H. P. (ed.) (1991) *Negotiation Analysis* (Ann Arbor: University of Michigan Press).

9.   The micropolitics of education: mapping the multiple dimensions of power relations in school polities

Betty Malen

'Micropolitics' is generally viewed as a new field of study. Since its conceptual boundaries and distinctive features are elusive and contested, this chapter adopts a working rather than a consensus definition of the field. Micropolitical perspectives address the overt and covert processes through which individuals and groups in an organization's immediate environment acquire and exercise power to promote and protect their interests (Ball 1987, Blase 1991b, Hoyle 1986). They emphasize the public and private transactions through which 'authorities' and 'partisans' manage conflict and meld consensus about the distribution of scarce but prized material and symbolic resources.[1] The processes through which actors broker roles, develop agreements and make decisions and the impact of these exchanges on the distribution of valued outcomes become the foci of study. Micropolitical models could be applied to units at any level of the system (e.g., Bacharach and Mundell 1993). Since the referent in education is often the school, this chapter provides an overview of the 'micropolitics' of schools, synthesizes select research, and identifies issues confronting the field.

Overview of the 'micropolitics' of schools

Elements of 'micropolitical' perspectives have been evolving for some time. These developments and other forces have rekindled interest in and spawned research on multiple dimensions of power and politics in schools (e.g., Marshall and Scribner 1991).

Conceptual developments: organizations as political entities

Various literatures have supported the notion that organizations are 'intrinsically political in that ways must be found to create order and direction among people with potentially diverse and conflicting interests' (Morgan 1986: 142). Moves from rational to natural and open systems models of organizations paralleled if not precipitated efforts to develop political models (Scott 1992). Criticisms of sociological and psychological explanations of organizational dynamics (e.g., the tendency to presume consensus or reify structure and personality) complemented, perhaps catalyzed efforts to develop political explanations (e.g., Bacharach and Lawler 1980, Hardy 1987, Pfeffer 1981). Studies of policy implementation discovered that the political bargaining endemic in enactment continues throughout implementation as 'street-level' service providers (Lipsky 1977) 're-make' policy (Bardack 1978, Berman and McLaughlin 1978, Weatherley and Lipsky 1977). Research on 'planned change' ventures identified organizational 'politics' as a factor shaping their outcomes[2] and corroborated claims that change processes are inescapably

0268–0939/94 $10·00 © 1994 Taylor & Francis Ltd.

political (Mangham 1979, Mann 1976). Treatises on administration also referred to political views of organizations and the 'politics' of management (e.g., Bowles 1989, Burns 1961, Campbell 1977–78).

These and other political conceptions of organizations echo and augment ideas evident in the sociology of education (e.g., Lortie 1969, Waller 1932) and the politics of education (Campbell *et al.* 1985, 1987, Iannaccone 1975, Sroufe 1977, Wirt and Kirst 1975). Scholars in these fields have long recognized that schools are mini political systems, nested in multi-level governmental structures, charged with salient public service responsibilities and dependent on diverse constituencies. Confronted with complex, competing demands, chronic resource shortages, unclear technologies, uncertain supports and value-laden issues, schools face difficult, divisive allocative choices. As in any polity, actors in schools manage the inherent conflict and make the distributional decisions through processes that pivot on power exercised in various ways and in various arenas. These processes are amenable to political analyses, but they have received limited examination, in part because 'politics' has been seen as an 'unprofessional' activity to be avoided, not an inevitable force to be addressed (Viteritti 1986). Simply put, the politics of schools has received more attention than the politics in schools.[3]

Empirical studies: power-relations emphasis

The literatures sketched above provide orienting frameworks for studying politics, but they have not yielded an over-arching theory of 'micropolitics' (Bacharach and Mundell 1993, Burlingame 1988, Townsend 1990). Since power is a central component of political analyses, much of the work seeks to map how power is acquired and exercised in schools. Since power is a 'primitive,' contested concept (Bacharach and Lawler 1980: 14), studies reflecting the 'cardinal assumption' that politics is about the acquisition and exercise of power draw on different definitions of power and its companion terms, authority, influence, and control (Geary 1992: 9).

To illustrate, some employ 'pluralist' views and concentrate on the overt manifestations of power evidenced by influence (or noninfluence) on salient, contentious decisions. Others draw on 'élitist' views that emphasize the more covert expressions of power apparent in the suppression of dissent, the confinement of agendas to 'safe' issues, the manipulation of symbols and the 'suffocation' of 'demands for change in the existing allocation of benefits and privileges' (Bacharach and Baratz 1970: 44). Still others draw on 'radical' (Lukes 1974) or 'critical' views. These delve into the more opaque if not invisible 'third face' of power and derive inferences on how power relations shape aspirations and define interests through subtle but presumably detectable processes of socialization/indoctrination that elude the awareness of the individuals who succumb to them but may be evident to the analyst who searches for them (Gaventa 1980, Lukes 1974). All these views of power have their advocates and critics (Geary 1992). All have made their way into studies of the 'micropolitics' of schools.

Spectrum of studies and focus of chapter: actors and arenas

Since research on site-level politics embodies different 'faces' of power, the literature is diverse. Studies span the space from neighborhood politics to classroom and corridor dynamics, move from organizational to interpersonal, at times intrapersonal units of analysis, encompass a maze of actors and address a range of topics (e.g., how political orientations are acquired; how political contests are conducted or averted; how politics impacts on careers, how it affects/reflects issues of race, class, and gender).[4] These works cannot be exhaustively reviewed in a single chapter or elegantly arrayed around a parsimonious theory. Thus this chapter synthesizes, in an illustrative versus exhaustive fashion, major findings on actor roles and relationships in select formal and informal arenas. It concentrates on professional–patron and principal–teacher interactions since those have received the most empirical attention. It is confined to prevalent themes and focuses on stakes (the what of politics, the content of contests), patterns (forms or 'styles of play'), strategies (how patterns are promulgated) and outcomes (impacts on participants and schools).

Professional–patron transactions in formal decision arenas

Principals, teachers, parents and community residents interact in formal arenas, such as program-specific advisory committees, school-wide advisory councils and school-based governing boards. These avenues for citizen involvement are rooted in ideals of local, democratic control; criticisms of closed, cumbersome 'bureaucratic' systems; and issues surrounding the equity and quality of educational opportunities. They have been created and re-created in response to federal and state laws as well as local pressures.[5] They have been examined through surveys of participant responses and case studies that depict decision dynamics and impacts.[6]

Stakes

Surveys and case studies indicate that professional–patron tensions pivot on who has the legitimate right to decide policy and whether the school has provided appropriate educational services. Generally speaking, principles and teachers subscribe to the tenet that professionals should make school policy and parents should endorse their decisions (e.g., Davies 1980, 1987, Mann 1974, Moles 1987). While parents express different preferences, they often accept that presumption (e.g., Davies 1981, Malen and Ogawa 1988).[7] But they also intermittently, at times vociferously challenge that premise, reassert their right to participate in decisions and place demands for reform on schools.[8] Parents tend to mobilize when events (e.g., desegregation plans, curricular changes, school closures) signal that discrepancies between community expectations and school operations exceed the 'zone of tolerance' (Iannaccone and Lutz 1970). Yet professionals recognize that parents could level criticisms that threaten the stability and legitimacy of the school at any time. Thus tensions regarding the role of parents in policy making, fears of 'intrusion' by 'outsiders' (Hempel 1986: 136), anxieties about the school's ability to withstand scrutiny and conflicts over divergent views of appropriate, equitable education are ever-present.

Prominent pattern

These sources of stress are typically managed through cordial, ceremonial exchanges that reflect and reinforce a traditional pattern of power wherein professionals, notably principals, control school policy, teachers control instruction, and parents provide support (e.g., Davies *et al.*, 1977, Jennings 1980, Malen *et al.* 1990). There are exceptions to this pattern, but it encapsulates the prominent themes in studies of (a) program-specific (e.g., Chapter I, Title I, special education) advisory committees where the final authority to make decisions is granted to the principal (e.g., Davies 1980, Fisher 1979, Shields and McLaughlin 1986) and (b) site-based decision-making bodies that broaden the jurisdiction of councils and suggest rhetorically, if not structurally, that parents are 'on parity' with professionals (e.g., Berman, Weiler Associates 1984, Bryk *et al.* 1993, Malen and Ogawa 1988). Since studies were open to if not guided by pluralistic views of power that assume partisans (parents) can influence authorities (principals), the signs of limited parent influence appear to be an attribute of professional–patron relations, not an artifact of a preemptive assumption that 'élites' inevitably control decisions.

The pattern of limited parental influence is evident in several ways. The parent-populated councils studied rarely address critical issues of budget, personnel and program (e.g., Davies 1980, Fisher 1979, Jennings 1980). Parents typically depict agendas as 'trivial' and identify issues they would prefer to discuss but are not able to raise (e.g., Davies 1987, Malen and Ogawa 1988, Mann 1974). In some cases parents 'provoke critical conversation about education practice, tracking, multiculturalism and racism and the splitting of vocational and academic curricula' (Fine 1993: 699) and voice other concerns. But parents also get 'stonewalled,' (Fine 1993: 699) or sanctioned when they do so (Malen and Ogawa 1988). As a result, councils rarely function as forums 'for meaningful discussion of significant issues and for educator–citizen problem-solving and power sharing' (Davies 1980: 63, Davies *et al.* 1977, Malen *et al.* 1990). They may operate, on occasion, as a caucus and influence policy at the district level (e.g., Iannaccone 1991, McLeese 1992). But at the building level the councils tend to be 'artificial bodies' whose existence [gauged by opportunities to influence policy] is 'more significant on paper than in practice' (Berman, Weiler Associates 1984: 166). However, the councils may be useful entities since they can 'create the impression' of parent involvement (Huguenin *et al.* 1979) and provide a means for contentious issues to be brought 'into a supportive structure under the control of district and school administrators' (Shields and McLaughlin 1986: 8). Since council membership does not, on demographic dimensions, reflect the economic, social, racial and ethnic composition of the school community, assessments generally conclude that councils are not representative bodies (Conway 1984, Malen *et al.* 1990, Mann 1975, 1977). Given these features, various parent councils have been characterized as 'proforma' units and public relations vehicles, not democratic decision-making bodies or policy-making entities (e.g., Davies 1987, Mann 1974, 1977, Popkewitz 1979). Thus parent influence in these arenas is more a goal to be pursued than a condition that has been realized, especially for low-income and minority populations (e.g., DeLacy 1992, Fine 1993).

Strategies

The above pattern is produced, in part, by the principal's capacity to preempt or curb parent voice. That is not to say principals are omnipotent actors. At times besieged by community pressures and constrained by community expectations, principals may be 'captives of their environments' (McPherson *et al.* 1975), powerless middle managers (e.g., Goldring 1993) caught in cross-currents of legislative mandates, district regulations, union contracts, constituency demands, teacher expectations, student pressures and their own convictions. Still, by virtue of their position as gatekeepers, principals can filter demands and affect deliberations in potent ways. They can leverage the composition of councils, an advantage that enables them to invite traditional supporters to be members, coopt vocal critics and 'socialize' parents into a supportive, at times submissive role (e.g., Goldring 1993, Malen *et al.* 1990). As the ones in charge of and accountable for the schools, principals have resources (e.g., stature, information, prerogatives) that can be used to control the agenda and 'create an aura of smooth sailing and mutual admiration that [leaves] policy determination intact in the principal's hands' (Mann 1974: 295).

The principal's ability to control decision processes and outcomes is augmented by teachers' willingness to align with the principal to keep major issues in the purview of professionals (e.g., Berman, Weiler Associates 1984, Mann 1974) and the ability to circuit contentious topics to private arenas such as a 'subcommittee' of principals and teachers (e.g., Malen and Ogawa 1988). The pattern is also the result of parents' reluctance to challenge the dynamics. For a mix of reasons such as deference to the expertise of professionals, limited information on actual school operations, 'serve and support' orientations, appreciation for being 'invited' to join the council, parents tend to be reticent partisans (e.g., Berman, Weiler Associates 1984, Chapman and Boyd 1986, Salisbury 1980). Further, norms surrounding professional prerogatives and harmonious interactions can mute discussion, muzzle conflict and maintain traditional patterns of power (e.g., Huguenin *et al.* 1979, Malen and Ogawa 1988). The often vague but narrow authority delegated and the modest supports (e.g., time, money) provided also restrict parents' incentives and opportunities for influence (e.g., Berman, Weiler Associates 1984, Fager 1993). Finally, the propensity for participation to be socially stratified suggests that matters of race, ethnicity and economics affect parent access to, hence prospects for influence in these arenas (e.g., Fine 1993, Malen and Ogawa 1988, Mann 1977).

Alternative styles of play

Whether efforts to lift the structural limits on parents' influence through policies that grant parents choice as well as voice will succeed remains an open, empirical question. Data on the impact of parent choice on the capacity to influence policy are thin, but they suggest the option to exit one school and enroll in another does not translate into influence on school policies, particularly for low-income and minority populations (e.g., Cohen and Farrar 1977, Kirp 1992, Moore and Davenport 1989). The most current, visible and studied experiment with school-based boards that significantly alter the balance of professional–parent power is under way in Chicago (Moore 1991). Even here, the pattern of limited parent influence is apparent on elected councils that give community members a numerical advantage, grant them the right to hire/fire the principal and otherwise expand their formal powers (Bryk *et al.* 1993, Wong 1994). There are sites where parents have become active through 'adversarial politics' and sites where principals have stimulated and/or parents have seized opportunities to move toward 'strong democracy' (Rollow and Bryk 1993b). But the incidence of 'maintenance politics' and 'consolidated principal power' suggests parents may be peripheral players at many sites. Although policy provisions redistribute formal powers in ways that should redound to the parents' advantage, factors that constrain parent influence may still be operative (Malen 1994b, Wong 1994).

Outcomes

Professional–patron interactions in these arenas beget mixed reviews. Across studies, professionals express appreciation for the support parents provide and concerns about the stress generated by the inclusion of constituency groups in decision areas and the apprehension that issues could get 'out of hand' (e.g., Chapman and Boyd 1986, Wiles 1974). Parents report gratitude for instrinsic rewards they procure (e.g., a sense of belonging) but frustration with 'token involvement.' Professionals and patrons indicate that interactions can foster understanding among individuals and groups and intensify tensions among them; they can foster support for and intensify dissatisfaction with the school (e.g., McLeese 1992).

While council interactions engender mixed sentiments, the evidence indicates that traditional professional–patron influence relationships are not fundamentally altered and may well be reinforced because the formal arenas operate as vehicles to air complaints and assuage concerns (Mann 1974, McLeese 1992); as forums to rally support for/reduce resistance to policies made elsewhere; as symbols of parents' right to voice in decisions and as signs that the organization recognizes that right. In these and other ways, the formal arenas can regulate conflict, reduce the likelihood that familiar influence relationship will be challenged and increase the chances that established relationships will be sustained (Malen *et al.* 1990, Popkewitz 1979). It is plausible that interactions could catalyze changes more commensurate with reform expectations (e.g., Moore 1991, Salisbury 1980), but the evidence to date suggests that maintenance patterns may be more probable.

The available evidence also indicates that council interactions have not substantially affected the quality and distribution of educational outcomes. The councils may stimulate adjustments, but these are often 'cosmetic' or 'cutback' in nature, in part because the tendency to revive parent councils during periods of intense fiscal strain puts the focus more on what to cut than how to improve (e.g., Fine 1993, Malen 1994a, McLeese 1992). That councils 'struggled to maintain their basic operations, few produced significant improvements' captures the conclusion of much of the research (Berman, Weiler Associates 1984: ii, Malen *et al.* 1990). Whether these tendencies will be reversed under 'empowered' councils created by the Chicago reform remains an open, empirical question. Various styles of play are evolving; signs of 'democratic politics' and organizational adjustments are surfacing in about a third of the elementary schools (Bryk *et al.* 1993). Ongoing research on how these developments affect the patterns of power and the quality/distribution of services should bolster our knowledge of (a) the conditions that precipitate different patterns of politics and (b) the connections between governance structures, political processes and education outcomes (Malen 1994b).

Professional–patron transactions in informal arenas

Professionals, parents and community residents interact through informal exchanges. These transactions have not received much empirical attention, but several studies address how principals manage exchanges with parents and publics.[9]

Stakes

The studies indicate that, as in any political exchange, the capacity of authorities or 'allocators' of resources to manage conflicts that arise as partisans or 'petitioners' make demands on the school is a central issue (Summerfield 1971). The amount of resources, the distribution of services, the dispersement of rewards or sanctions and the meaning of propriety and fairness are also matters of concern and contest (e.g., Blase 1989, 1991a).

Prominent patterns and strategies

The actual or anticipated tensions are mediated largely by the principal through strategies to avert or contain conflict (e.g., Hempel 1986, Wiles 1974). The principal, structurally situated as 'gatekeeper,' is positioned to buffer the school from external influences and to filter, forward or forestall demands. That is not to say principals act unilaterally or uniformly. Different communities engender different dynamics (e.g., Bryk *et al.* 1993) but in all cases principals surface as key actors whose primary political function is to 'minimize conflict between the community and the neighborhood school' (Summerfield 1971: 93). Their strategies have been most fully articulated by Summerfield (1971). Likening 'neighborhood based' politics to interest group politics, Summerfield argues that principals recognize the 'potential power of the public and its groups' and use four major situationally derived strategies to contain conflict and procure deference (1971: 4).

First, principals act as leaders. They seek resources from central authorities and use acquisitions to reinforce community support and redirect conflict toward central rather than school authorities. Parents and publics defer to school authorities in exchange for their signs of hard work or their success in securing resources. Second, principals behave as passivists. Amidst the presence of need but in the absence of a community impulse for action, passivists do not press for resources. This strategy avoids conflict and procures deference through limited knowledge of unmet needs. Third, principals act as symbol managers who lack the community consensus to mount resource campaigns and rely on 'symbolic reassurances' to 'assuage parental anxiety' (Summerfield 1971: 87). Deference flows from a 'denial of real problems' (Summerfield 1971: 95). Finally, principals act as 'nice neutrals' who occasionally call on the community to petition authorities but typically operate so that discontent is not aroused. Deference is secured by sustaining the perception that the school is doing well.

More recent works document additional actions that principals take to manage conflict with parents or community groups. Principals may selectively enforce discipline policies to 'avoid direct confrontation with outsiders,' notably parents who might expand the scope of conflict by involving the board, the courts or others (Corbett 1991: 94). This strategy accommodates some parents but, from the teachers' standpoint, undermines their authority and erodes the integrity of instruction, the consistency of discipline and the fairness of rules. Under pressure from high-status parents, principals learn to 'finesse requirements, to "wink" at improprieties and develop "live" and "let live" agreements' (Hempel 1986: 27). They also press teachers to alter grades, modify class content, adjust homework, or grant 'favored status' to students in particular programs (Blase 1988). These actions suggest that parents influence through private but poignant exchanges. They indicate that principals are not simply buffers that insulate the school from outside forces, but arbiters of disputes, negotiators of private compacts and conduits for parent influence on programs and practices.

Outcomes

Data on the frequency and impact of these strategies are limited but several observations can be made. The broad strategies arrayed by Summerfield seek to minimize conflict, not equalize power. They enable principals to regulate whether and how patrons are engaged in efforts to secure resources and address problems. Since principals who act as spokespersons for neighborhoods are better positioned to secure allocations, patterns of neighborhood politics may have consequential effects on the distribution of resources to schools, if not the distribution of power among professionals and patrons at schools. Whether individualized private agreements fundamentally alter the professional–patron power equation or intermittently moderate it, select parents exert influence on important aspects of the school. The cases suggest micro-sized deals have 'more than "micro-sized" effects' (Corbett 1991: 87). They may undermine teachers' efficacy, lower morale, intensify stress and erode commitment, particularly when the deals violate cherished norms of propriety and fairness (e.g., Blase 1988, Corbett 1991). They may affect student attitudes and student access to programs. They may placate the demands of select parents in ways that are unfair to less vocal or powerful constituencies. They may 'maintain smooth operations by deflecting fundamental challenges to those operations' (Bryk *et al.* 1993: 7) and thereby reinforce existing patterns of power and privilege (e.g., Rollow and Bryk 1993a).

Principal–teacher transactions in formal decision arenas

Principals and teachers interact in a variety of formal arenas like faculty meetings, site councils and improvement teams. These shared decision-making forums have been part of school governance structures for decades and have been examined through surveys of participants' responses and case studies of decision dynamics.[10]

Stakes

Surveys and case studies indicate that transactions center on tensions surrounding who has the legitimate right to make decisions in particular territorial (e.g., schoolwide, classroom bounded) and topical domains (e.g., budget, personnel, curriculum, instruction). Principals and teachers are not monolithic groups. Principals are inclined, however, to see certain topics (e.g., budget, personnel) and territories (e.g., schoolwide policy) as administrative prerogatives in part because authority over these matters has been vested in the principal's office (e.g., Conley 1991, Weiss 1993). Teachers are interested in acquiring influence in these domains and protecting their authority over other areas, notably classroom instruction (e.g., Bacharach *et al.* 1986, Rowan 1990). Since topics and terrains of interest to both parties fall in 'contested zones' (Hanson 1981), principals and teachers negotiate who should have influence on particular decisions. They constantly broker the boundaries of their respective 'spheres of influence' (Hanson 1981). In so doing, they negotiate what if any changes will be permitted in school policies/practices and what (or whose) conception of the school will prevail (e.g., Ball 1987).

Prominent pattern

Principals and teachers tend to manage these tensions through polite exchanges that mirror and maintain the traditional pattern of power wherein principals retain control over school-level policy decisions and teachers retain discretion over classroom-level practices (e.g., Duke *et al.* 1981, Malen and Ogawa 1988). Corwin captures the essence of the transaction: 'Teachers purchase discretion within classrooms by relinquishing their opportunity to influence school policy' (1981: 276). While there are exceptions, this pattern captures the dominant themes in research on principal–teacher interactions in formal arenas. Since studies have embraced pluralistic views of power that assume partisans (teachers) can influence authorities (principals), the maintenance of conventional relations is an unanticipated finding. It is apparent, however, in that major topics (e.g., budget, personnel) get mentioned but are not regularly addressed (e.g., Malen *et al.* 1990, Weiss 1993). Teachers view agenda items as trivial and tangential, as 'small decisions' (Mauriel and Lindquist 1989) and depict options for input as token gestures that rubberstamp decisions already made (e.g., Johnson 1989, 1990) as 'pseudo-participation' (Ball 1987).

Strategies

The above pattern is produced in part by the principals' capacity to control decision processes. Given their position, principals are inclined and equipped to protect their managerial prerogatives by controlling the agenda content, meeting format and information flow (e.g., Ball 1987, Corcoran 1987, Gronn 1984, Malen and Ogawa 1988, Mann 1974). In some cases they recruit supportive teachers as council members, form coalitions with teacher allies or overturn troubling decisions by not implementing them (e.g., Bryk *et al.* 1993, Hanson 1981, McLeese 1992, Weiss and Cambone 1993). But the pattern is also the result of teachers' reluctance to challenge the principal's definition of the situation, a reluctance shaped by many factors but rooted in the fear of social and professional sanctions (e.g., being cast as a troublemaker, a malcontent) that may be applied by principals and peers alike (e.g., Corcoran 1987, Duke *et al.* 1981). Such sanctions (actual or anticipated) are potent because they can taint reputations, jeopardize advancements and denigrate persons (e.g., Ball 1987, Weiss *et al.* 1992). Further, they are perpetually reinforced by ingrained norms (e.g., harmony, civility) that can mute dissent, minimize conflict, and maintain stability (Malen and Ogawa 1988). Finally, the pattern may be related to the limited authority delegated to sites and the modest supports (e.g., time, money) provided (e.g., David and Peterson 1984, Fish 1994, Stevenson and Pellicer 1992).

Alternative styles of play

The prominent pattern is not universal. In some settings teachers report that they influence school policy decisions (e.g., Bryk *et al.* 1993, Carnoy and MacDonell 1990, DeLacy 1990, 1992, Smylie 1994, White 1992). Even in these sites, however, their influence is contingent on the principals' willingness to share power (e.g., Smylie 1992, Weiss and Cambone 1993) and the administrators' (district and building) propensity to disperse responsibility for decisions on contentious issues like budget cuts (e.g., Corcoran 1987, McLeese 1992, Smylie and Tuermer 1992). Thus it is, as Weiss notes, 'hard to avoid the sense . . . teachers are being coopted,' rather than 'empowered' (1993: 89).

To be sure, there are cases in which teachers mobilize coalitions that effectively check the principal's capacity to control policy decisions or override principal-supported or principal-instigated initiatives (e.g., Hanson 1981, Muncey and McQuillan 1992, Weiss and Cambrone 1993). There are instances where teachers voluntarily align with the principal's agenda around sets of shared interests (e.g., Goldman et al. 1993). There may be instances in which teachers initiate as well as 'implement' policy forged elsewhere. But such episodic interruptions do not markedly alter the dominant pattern wherein site actors 'make' policy within fairly narrow parameters set by the broader system (Fish 1994, Hargreaves 1991), and, within these circumscribed boundaries, principals retain control of schoolwide decisions and teachers retain control of classroom-specific decisions (e.g., Ball 1987, McLeese 1992). Despite the recurrent infusion of 'new' structures to bolster teachers' influence, the traditional pattern persists, even in sites with extensive experience in and commitment to shared decisions (Lichtenstein et al. 1992, Malen 1993, Weiss 1993).

Outcomes

Principal–teacher interactions in these formal arenas engender consequential effects. The studies indicate that interactions can augment or erode teacher efficacy, kindle or curb their willingness to participate in decisions, direct or divert attention to the instructional component of schools, contribute to or detract from satisfaction with work and commitment to the organization (e.g., Chapman and Boyd 1986, Conway 1984, Rowan 1990, Shavelson 1981, Weiss et al. 1992). By and large, teachers may be initially enthusiastic. But they get weary and wary. They get exhausted by the demands and become skeptical of the prospects for meaningful influence and suspicious of the requests for their involvement (e.g., Johnson 1990, Malen et al. 1990). Principals are also apprehensive. While some endorse the broad ideal, they too get exhausted and frustrated by the demands of participatory processes and struggle to retain control over issues perceived to fall within their topical and territorial domains (e.g., Bennett et al. 1992, McLeese 1992). While incremental changes may be brokered, major changes in actor roles/relationships or school operations/outcomes are not apparent in many of the settings studied (e.g., Berman, Weiler Associates 1984, Bryk et al. 1993, Malen 1993, McLeese 1992, Weiss and Cambone 1993). The link between teacher participation in school decision-making and the alteration or improvement of classroom instruction is difficult to establish, but this 'classroom connection' (DeLacy 1990, 1992), is not apparent in most sites studied (e.g., Fullan 1993, Smylie 1994).

An analysis of reasons for these outcomes goes beyond the scope of this chapter. What is more relevant here is that, despite mixed responses, there are strong indications that these structures serve important political functions, comparable to those of parent councils. They operate as vehicles to air complaints and assuage concerns (e.g., Ball 1987, Mann 1974, McLeese 1992); as forums to rally support for policies made elsewhere (e.g., Hanson 1991, Malen 1993, 1994a); as symbols of teachers' right to a voice in decisions and as signs that the organization recognizes that right. In these and other ways, interactions in formal decision arenas can reduce the likelihood that traditional patterns of influence will be challenged and increase the likelihood that established patterns will be sustained. It is possible to engender effects more consistent with reform expectations (e.g., Goldman et al. 1993), but the evidence suggests maintenance patterns may be more probable.

Principal–teacher transactions in informal arenas

Principals and teachers interact through a variety of informal contacts that occur in schools. Many writings allude to the political character of these conversations, but few array the political dynamics of principal–teacher exchanges in these 'backstage' arenas.[11] The relatively recent efforts to examine these exchanges from explicitly political perspectives are derived from ethnographic studies of 'everyday interactions' in schools, case accounts of 'change' (leadership succession, program adjustments, career advancements) and reports of political practices in schools (e.g., Ball 1987, Blase 1989, 1991a, 1991c).

Stakes

These works suggest that teachers' professional security and integrity are at risk. Simply said, teachers are vulnerable to the criticisms of principals, peers, parents and students (e.g., Blase 1988). Thus they insulate themselves from the pressure and pain of interactions that can damage their reputations, diminish the quality of their work life and disrupt their ability to carry out their responsibilities in ways congruent with their views and values. They are cognizant of the principal's power to support or sanction their actions. Principals can allocate discretionary funds, adjust assignments, influence evaluations and promotions and otherwise affect teachers' well-being. The vulnerability of their position, the uncertainty of the principal's response, and the perception that principals (and others) practice 'favoritism' (Blase 1988) in the distribution of rewards and reprisals make 'everyday interactions' consequential conversations for teachers.

From the vantage point of principals, what is at stake is their capacity to exercise control over and engender commitment to school policies/practices (Ball 1987: 84). As the one ultimately responsible for the organization, the principal is in a vulnerable as well as a pivotal position. Their stature in the system, their ability to carry out their duties in ways congruent with their views and values are on trial (e.g., Bridges 1970, Hanson 1981). Thus both principals and teachers seek to protect their well-being, preserve their respective 'spheres of influence' (Hanson 1981), and procure a climate of harmony in the school and an image of legitimacy in the community (e.g., Ball 1987, Blase 1991a).

Prominent pattern and strategies, alternative styles of play

Teachers and principals tend to manage these interrelated interests and uncertainties through strategies that maintain familiar role boundaries and established organizational practices. While there are exceptions, the maintenance pattern is prominent in the studies and is produced through strategies indicative of, but not confined to, those associated with the subtle faces of power.

Teachers' strategies: Teachers, generally speaking, deploy a set of protective strategies with principals. Following Blase's (1988) typology of learned responses, teachers acquiesce. They accede to directives and requests initiated or supported by the principal even though the actions sought and secured may violate their views of ethical practice. They conform to rules and norms in part because they concur with them and in part because they calculate the costs of questioning them (e.g., alienating the principal, evoking sanctions from the principal). They circumvent (e.g., quietly disregard directives), sabotage (surreptitiously undermine the principal's credibility through gossip, innuendo), and adopt other forms of 'passive resistance' (Blase 1988). They ignore requests or create ingenious responses that blunt or 'veto' directives primarily because the requests violate their views and values.[12]

Teachers also deploy a set of promotional strategies, apparent in efforts to ingratiate by flattering or appeasing the principal, to voice positions in diplomatic ways, and, in rare instances, to confront directly the issues at hand (Blase 1988, 1989). Moreover, teachers deploy what might be termed 'preparatory strategies' to accumulate resources for influence (e.g., do extra work, secure visibility for work, develop expertise for more informed advocacy). On occasion they use these and other resources to mobilize factions and forge coalitions, to persuade and pressure principals, to advocate and advance counter-proposals and contain or curtail the principal's initiatives (Blase 1991a, Hanson 1981). But these more assertive, collective strategies appear to interrupt, not overturn, the reliance on protective, individualized strategies of 'acquiescence', 'avoidance', and 'accommodation' (Blase 1991a: 363). As earlier noted, teachers appear to be 'reluctant partisans' for many reasons, including the fear of censure and sanction that may be imposed by principals and peers (Blase 1988). While teachers' strategies to influence principals will vary with the issue, the 'style' of the principal and the strength of alignments, 'protective' strategies are pronounced in the sites studied.

Principals' strategies: Principals, generally speaking, employ an array of control strategies with teachers. Principals may learn through induction processes to avoid or contain conflict (Marshall and Mitchell 1991). Be it for these or other reasons including a genuine desire to advance their view of what is best for the school (Hanson 1981), principals may confine agendas to safe issues, 'consult' with teachers selectively or ritualistically to preempt or coopt resistance (Ball 1987, Blase 1988); broker agreements or grant favors that dispel opposition, engender indebtedness, or 'induce loyalty' (Blase 1988: 163, Ball 1987). They may 'stifle talk' (Ball 1987: 110) or 'silence' teachers (Anderson 1991: 136) through rhetorical devices that define discussion as 'subversive' action, break faculties into fragmented factions and/or 'stigmatize and isolate opponents' (Ball 1987: 137). They may also manipulate the information flow and 'manage the meanings' ascribed to actions (Anderson 1991: 122). While these and other strategies do not go uncontested, and while their form and frequency may vary with the 'style' of the principal and the issue at hand, the stock of strategies suggests principals can effectively 'block, stifle, dissuade or ignore groups in school who advocate change' (Ball 1987: 79, Berman and McLaughlin 1978, McLaughlin 1987). Principals can and do deploy other strategies. They can coalesce (as well as control) teachers through moral suasion (e.g., Greenfield 1991); they can 'bring out the best in teachers' with strategies (e.g., recognition, praise) that may engender a different style of play, if not a different balance of power (Blase and Kirby 1992, Greenfield 1991). Still, the prominent pattern in studies of 'everday interactions' is marked by strategies that cumulatively and synergistically sustain conventional roles and practices.

While this prominent pattern is not universal, it appears durable. Simply put, various efforts to redefine roles (e.g., to create teacher leader posts) or otherwise 'change' schools have a tough time taking hold in part because the subtle exchanges between principals and teachers translate 'new' ventures into conventional arrangements (e.g., Little 1990, Malen 1993). Case studies of these initiatives indicate that teachers and principals broker provisions to avert the social sanctions and contentious exchanges that can accompany the redefinition of roles and relationships in organizations (e.g., Malen and Hart 1987a,

1987b). Accounts of other change ventures suggest that various initiatives are subject to ongoing political negotiations that often marginalize the impact of the planned change (e.g., McLaughlin 1987, Muncey and McQuillan 1992, Weatherley and Lipsky 1977). To be clear, these outcomes ought not be construed as indictments of educators' character, competence or courage but as responses to a host of pressures that evoke survival tactics and protective posturings.[13]

Outcomes

These informal exchanges appear to have contingent effects on teachers. When principals rely on 'control' strategies, teachers may become 'less engaged, less motivated and less committed' (Blase 1988: 124); when principals exert pressure and teachers accede to demands that violate their values, teachers may become more stressed and more inclined to exit the occupation (Blase 1988: 154). Conversely, other strategies may enhance teachers' morale and commitment to the school (Blase and Kirby 1992).

The 'politics' of principal–teacher interactions is a major source of stress for principals (Bridges 1970, Marshall and Mitchell 1991) and a force that has organizational effects. As in any polity, informal exchanges between principals and teachers 'bring an acceptable degree of order and stability' to the school (Hanson 1981: 266) and tend to make change, for better or worse, an incremental if not an incidental outcome (e.g., Ball 1987).

Observations on the 'micropolitics' of schools

This illustrative look at research on the 'micropolitics' of schools supports three general observations.

First, the 'micropolitics' of schools is a disparate field. Its conceptual boundaries and distinctive features await definition. The spectrum of work suggests the terrain is elusive, at times all-inclusive. By some accounts, any human interaction is a political interaction, every conversation is a caucus, every move is a maneuver that somehow affects/reflects the 'politics' of the site. Clearly, conversations can be sources of political intelligence, arenas of political negotiation, conduits for political connections and avenues of political influence. Actions can have political consequences whether they are meant to or not. But, if micropolitics is much ado about everything, is it much ado about anything? What is the essence of micropolitics? How does it differ from macropolitics? Or does it? Is 'micropolitics' defined by the size of the arena? The level of the system? The unit of analysis, which in works reviewed here stretches from the psyche of the individual to the polity of the organization? Is it defined by the style of play? Is micropolitics essentially the politics of 'privatization' (Schattschneider 1960), a politics that confines the scope of conflict to safe issues, restricts the game to insider exchanges and puts the emphasis on the acquisition of acquiescence? Is micropolitics a limited set of games in the broader 'ecology of games' (Firestone 1989)? In short, what are the conceptual parameters and distinguishing features of the field? Fortunately, efforts to address these issues are under way (e.g., Bacharach and Mundell 1993, Ball 1987, Rollow and Bryk 1993b). While these developments have not yet yielded a consensus definition, they have directed attention to the need for conceptual clarification and provided frameworks to guide research and ground understandings.

Second, if one accepts the cardinal assumption that politics is in large measure about the acquisition and exercise of power in a polity, an assumption upheld in this chapter, then one challenge confronting the field is the fortification of a research base for whatever 'face' of power is the focus of study. Much of the research reviewed here concentrates on the first face and examines how power is manifest in influence (or non-influence) on decisions. The prominent patterns indicate that neither parents nor teachers exercise significant influence on significant issues in formal decision arenas. While there are exceptions, the tendency for processes to maintain familiar power configurations is pronounced. Our capacity to account for these dynamics is constrained by the lack of longitudinal and comparative data. Further, our capacity to grapple with the consequences of different patterns of politics is constrained by the limited attention to connections between political processes and education outcomes. More robust designs that probe actor relations, the conditions that produce, perpetuate or precipitate shifts in patterns of politics and the consequences of these styles of play for the distribution of valued outcomes would bolster our ability to interpret the politics in schools. Moreover, much of the research reviewed here underscores the importance of examining additional 'faces' of power. Actor roles, strategies and interactions in formal and informal arenas resemble the subtle, covert processes associated with the second and third faces of power. A problem here, however, is that these processes are exceedingly difficult to untangle. Efforts to get at them encounter methodological minefields that scholars have tried to attend, largely through applications of anthropological methods that require extensive and intensive time in the setting under study. The most credible research in this tradition has been based on years of time in the sites of study (e.g., Gaventa 1980). The extensive ethnographic work required to illuminate these subtle processes would be a substantial investment but an important, if not essential complement to the research presently available.

Third, if power is a key component of politics, then attending to the relationship between the three faces of power constitutes another challenge. One way to view the relationship is to see the faces as complementary. To illustrate, the first face concentrates on overt political action in decision arenas, on how power is activated and exercised. The second and third faces uncover the subtle precursors of political action such as how political orientations are formed, political efficacy is acquired, power resources are accumulated, public issues are defined and how broad structures as well as actor strategies converge to regulate the flow of influence. Taken together the three faces give a fuller understanding of political processes.[14] Recognizing that there are other ways to attend to the multidimensional nature of power, the point to be made is that diverse efforts to map power relations in schools are important steps. But these efforts might yield a more comprehensive, coherent account of politics if the 'faces' of power were integrated more effectively.

These three observations could be made about the study of politics writ large, not just the study of 'micropolitics' per se. This chapter leaves to others debates about the validity and virtue of even making a distinction between 'micro' and 'macro' politics. What seems crucial is that we understand politics as a force, for good or ill, and work to develop a keener understanding of the complexities and the consequences of power relations and political processes. The school is certainly a sensible place to anchor effort. It is an institution for political socialization, an object of political contest and an arena of political negotiation. It is surely a polity the warrants our fervent attention and our finest analysis.

Notes

1. Following Gamson (1968), authorities are organizational actors with the formal power to make binding decisions. Partisans are unofficial actors who might influence authorities. While authorities may influence partisans and other authorities, the key distinction is their formal power to make binding decisions.

2. While studies often cast politics as an unanticipated barrier to the consonant installation of proposed changes or as a peripheral problem to be averted by training in 'human relations' (Bolman and Deal 1991), they have documented the presence of 'politics' in schools.

3. Historians, philosophers, sociologists, anthropologists, critical theorists and others have addressed politically consequential aspects of schools (e.g., their social-civic purposes; the impact of governmental actions, public pressures and community mores on schools; the accommodations made to maintain the appearance of orderly, legitimate operations in schools). These works reveal the political character of schools and provide general profiles of power relations in schools, but they do not focus on the processes through which proximate actors (i.e., principals, teachers, parents, students) exercise power, or the manner in which contexts mediate their ability to influence policy or practice.

4. Accounts of dyads of master–mentor teachers, coalitions of school–community groups or partnerships of various sorts illustrate the maze of subsystems addressed (e.g., Baker 1994, Conley et al. in press, Heckman et al. 1994). Studies of teachers as 'policy brokers' in the classroom (e.g., Schwille et al. 1980, 1983) and accounts of classroom interactions indicate that 'authorities' (teachers) and 'partisans' (students) develop 'treaties' that have both attitudinal and instructional consequences (Powell et al. 1985, see also, McNeil 1986, Sedlak et al. 1986). These and other works suggest the classroom may be a telling 'micropolitical' arena (e.g., Bloome and Willet 1991, Elmore 1987, Guttman 1989). Cases of adult–student or peer-group interactions in informal arenas may illuminate school politics, particularly political socialization processes (e.g., Blase 1991c, Opotow 1991). Beyond embracing multiple arenas, studies concentrate on different aspects of power and politics such as the interests and ideologies of actors, their 'assumptive worlds,' or perceptual screens (e.g., Marshall and Mitchell 1991); their 'logics of action,' their sets of assumptions about means–end relationships (e.g., Bacharach and Mundell 1993), even their 'inner struggles' for 'identity in institutions that have the power to legitimate or delegitimate their voices' (Anderson and Herr 1993: 59). Some emphasize the strategy of ideological control, the management of meanings and myths and the suppression of dissent (e.g., Anderson 1990); the 'politics' surrounding a particular policy action (e.g., Ball and Bowe 1991), or organizational phenomenon (e.g., leadership succession; personnel selection), or the manner in which issues of race and gender are evident in and influenced by the micropolitics of schools (e.g., Ball 1987, Marshall 1993).

5. For an overview of parent involvement and community control movements, see: Campbell et al. (1987), McLeese (1992), Reed and Mitchell (1975), Rollow and Bryk (1993b), Tyack (1981).

6. Studies have been conducted by parent advocacy groups, research institutes, community activist organizations and professors. The data are plentiful but problematic. While survey instruments can elicit relevant data on compliance items (e.g., whether councils meet) and secure global assessments of participants' involvement in school affairs, they do not generate the detailed depictions of decision-making dynamics needed to determine whether 'involvement' at various stages of the process translates into meaningful or meager influence on major or minor issues. Unchecked by the corroborating (or correcting) patterns derived from depictions of influence processes, general assessments may cloud more than clarify conclusions about the constellations of power and the exercise of influence (Malen 1994b). Unfortunately, there are few systematic studies of decision-making dynamics, few comparative cases and few longitudinal examinations (e.g., McLeese 1992).

7. Studies conducted over the past two decades document that parents want greater influence on a broader band of issues than educators deem appropriate. Areas of contest include discipline, curriculum, program, finance and personnel selection and evaluation. For a review, see McLeese (1992).

8. As Drachler described it, dormant forces erupt, parents activate and align with community groups, and, if such mobilization escalates, the 'neighborhood school [can become] the people's city hall' (1977: 199), the object and arena of intense conflict that can polarize, at times paralyze the system. Not all eruptions take on 'adversarial' forms (Bryk et al. 1993). Community activism may be instigated or embraced by professionals and operate to alter and improve the system. Still, community mobilization poses a challenge professionals generally seek to avert or contain.

9. The ways parents, residents, local mores and other community forces shape the politics surrounding schools have been examined primarily from the vantage point of the district (e.g., Bacharach and

Mitchell 1981, Boyd 1976). Studies of school-based dynamics are quite rare (e.g., Rollow and Bryk 1993b, Wiles 1974).

10. The sheer volume of research on teacher participation in decision-making precludes thorough citation here. For reviews, see Conway (1984), Conley (1991), Rowan (1990). Much of the survey research focuses more on psychological dimensions (e.g., how participants feel about involvement) than political dimensions (e.g., whether participants influence decision outcomes). Thus survey data are used to inform judgements about expectations for participation and general responses to it. Case studies of teacher influence on decisions focus on a small number of schools in a particular context at a point in time. Most studies cited here were conducted in US schools, but some (Chapman and Boyd 1986, Ball 1987, Gronn 1984) were conducted in Australia and Britain. Since US studies often sample 'exemplar' sites, the representativeness of cases is questionable. Since there are few longitudinal or comparative studies, the ability to 'explain' patterns is limited.

11. Studies of policy implementation and planned change acknowledge 'political problems,' but rarely analyze political dynamics. Studies that focus on principal–teacher political dynamics were done in schools located in the USA and Britain.

12. As Hanson described it, 'Many teachers were magnificent in making it appear as though they were in complete support of an administrator's formal or informal intervention while all the time they were ignoring its every intent. . . . They genuinely saw themselves as guardians of the classroom and had to hold the line against what they considered to be fads . . . "gimmicks" . . . ' (1981: 269).

13. This contextualized interpretation is developed in a number of writings. See for example, Ball (1987), Blase (1988), Malen and Hart (1987b), Sarason (1990), Weatherley and Lipsky (1977).

14. Such an orientation has been carefully articulated by Clegg in his treatment of 'circuits of power' and his discussion of how this conceptualization enables one to 'demonstrate how networks of interests are actually constituted and reproduced through conscious strategies and unwitting practices' (Clegg 1989).

References

ANDERSON, G. (1991) Cognitive politics of principals and teachers: ideological control in an elementary school, in J. Blase (ed.) *The Politics of Life in Schools: Power, Conflict, and Cooperation* (Newbury Park: Sage), 120–138.

ANDERSON, G. L. (1990) Toward a critical constructivist approach to school administration: invisibility, legitimation, and the study of nonevents, *Educational Administration Quarterly*, 26(1): 38–59.

ANDERSON, G. and HERR, K. (1993) The micro-politics of student voices: moving from diversity of bodies to diversity of voices in schools, in C. Marshall (ed.) *The New Politics of Race and Gender* (New York: Falmer Press), 58–68.

BACHARACH, P. and BARATZ, M. S. (1970) *Power and Poverty* (New York: Oxford University Press).

BACHARACH, S. B. (1981) Organizational and political dimensions for research on school district governance and administration, in S. B. Bacharach (ed.) *Organizational Behavior in Schools and School Districts* (New York: Praeger), 3–4.

BACHARACH, S. B., BAUER, S. and SHEDD, J. B. (1986) The work environment and school reform, *Teachers College Record*, 88: 241–256.

BACHARACH, S. B. and LAWLER, E. J. (1980) *Power and Politics in Organizations* (San Francisco: Jossey-Bass).

BACHARACH, S. B. and MITCHELL, S. M. (1981) Critical variables in the formation and maintenance of consensus in school districts, *Educational Administration Quarterly*, 17: 74–97.

BACHARACH, S. B. and MUNDELL, B. L. (1993) Organizational politics in schools: micro, macro, and logics of action, *Educational Administration Quarterly*, 29(4): 423–452.

BAKER, L. M. (1994) The politics of collaboration: how an educational partnership works, paper prepared for the annual meeting of the American Educational Research Association, New Orleans, LA, April 4–8.

BALL, S. J. (1987) *The Micro-politics of the School: Towards a Theory of School Organization* (London: Methuen).

BALL, S. J. and BOWE, R. (1991) Micropolitics of radical change: budgets, management and control in British schools, in J. Blase (ed.) *The Politics of Life in Schools: Power, Conflict, and Cooperation* (Newbury Park: Sage), 18–45.

BARDACK, E. (1978) *Implementation Game* (Cambridge, MA: MIT Press).

BENNETT, A. L., BRYK, A. S., EASTON, J. Q., KERBOW, D., LUPPESCU, S. and SEBRING, P. A. (1992) *Charting Reform: The Principals' Perspective*, Report on a survey of Chicago Public School principals (Chicago: Consortium on Chicago School Research).

BERMAN, P. and McLAUGHLIN, M. (1978) *Federal Programs Supporting Educational Change, VIII* (Santa Monica: Rand).

BERMAN, WEILER ASSOCIATES (1984) *Improving School Improvement: A Policy Evaluation of the California School Improvement Program* (Berkeley, CA: Berman, Weiler Associates).

BLASE, J. (1988) The politics of favoritism: a qualitative analysis of the teachers' perspective, *Educational Administration Quarterly*, 24: 152–177.

BLASE, J. (1989) Teachers' political orientation vis-à-vis the principal; the micropolitics of the school, in J. Hannaway and R. Crowson (eds) *The Politics of Reforming School Administration* (New York: Falmer Press), 113–126.

BLASE, J. (1991a) The micropolitical orientation of teachers toward closed school principals, *Education and Urban Society*, 27(4): 356–378.

BLASE, J. (1991b) The micropolitical perspective, in J. Blase (ed.) *The Politics of Life in Schools: Power, Conflict, and Cooperation* (Newbury Park: Sage), 1–18.

BLASE, J. (1991c) Everyday political perspectives of teachers toward students: the dynamics of diplomacy, in J. Blase (ed.) *The Politics of Life in Schools: Power, Conflict and Cooperation* (Newbury Park: Sage), 185–206.

BLASE, J. and KIRBY, P. C. (1992) *Bringing Out the Best in Teachers: What Effective Principals Do* (Newbury Park: Corwin Press).

BLOOME, D. and WILLETT, J. (1991) Toward a micropolitics of classroom interaction, in J. Blase (ed.) *The Politics of Life in Schools: Power, Conflict and Cooperation* (Newbury Park: Sage), 207–236.

BOLMAN, L. G. and DEAL, T. E. (1991) *Reframing Organizations* (San Francisco: Jossey-Bass).

BOWLES, B. D. (1989) Gaining support for change: the politics of strategic leadership, in J. J. Mauriel, *Strategic Leadership for Schools* (San Francisco: Jossey-Bass), 163–210.

BOYD, W. L. (1976) The public, the professionals and educational policy making: who governs?, *Teachers College Record*, 77: 539–577.

BRIDGES, E. M. (1970) Administrative man: origin or pawn in decisionmaking?, *Educational Administration Quarterly*, 6: 7–24.

BRYK, A. S., EASTON, J. Q., KERBOW, D., ROLLOW, S. G. and SEBRING, P. A. (1993) *A View from the Elementary Schools: The State of Reform in Chicago* (Chicago, IL: Consortium on Chicago School Research).

BURLINGAME, M. (1988) Review of *The Micropolitics of the School: Towards a Theory of School Organization*, *Journal of Curriculum Studies*, 29: 281–283.

BURNS, T. (1961) Micropolitics: mechanisms of institutional change, *Administrative Science Quarterly*, 6: 257–281.

CAMPBELL, R. F. (1977–78) A history of administrative thought, *Administrator's Notebook*, 26(3): 1–4.

CAMPBELL, R. F., CUNNINGHAM, L. L., NYSTRAND, R. O. and USDAN, M. D. (1985) *The Organization and Control of American Schools*, 5th edn (Columbus, OH: Charles E. Merrill).

CAMPBELL, R. F., FLEMING, T., NEWELL, L. J. and BENNION, J. W. (1987) *A History of Thought and Practice in Educational Administration* (New York: Teachers College Press).

CARNOY, M. and MACDONELL, J. (1990) School district restructuring in Santa Fe, New Mexico, *Educational Policy*, 4: 49–64.

CHAPMAN, J. and BOYD, W. L. (1986) Decentralization, devolution, and the school principal: Australian lessons on statewide educational reform, *Educational Administration Quarterly*, 22: 28–58.

CLEGG, S. R. (1989) *Frameworks of Power* (Newbury Park: Sage).

COHEN, D. K. and FARRAR, E. (1977) Power to the parents? The story of education vouchers, *Public Interest*, 72–97.

CONLEY, S. (1991) Review of research on teacher participation in school decision making, in G. Grant (ed.) *Review of Research in Education* (Washington, DC: American Educational Research Association), 225–265.

CONLEY, S., BAS-ISSAC, E. and SCULL, R. (in press) Teacher mentoring and peer coaching: a micropolitics interpretation, *Journal of Personnel Evaluation in Education*.

CONWAY, J. (1984) The myth, mystery and mastery of participative decision making in education, *Educational Administration Quarterly*, 20: 11–40.

CORBETT, H. D. (1991) Community influence and school micropolitics: a case example, in J. Blase (ed.) *The Politics of Life in Schools: Power, Conflict and Cooperation* (Newbury Park, CA: Sage), 73–95.

CORCORAN, T. B. (1987) *Teacher participation in public school decision-making: a discussion paper*, paper prepared for the Work in America Institute.

CORWIN, R. (1981) Patterns of organizational control and teacher militancy: theoretical continuities in the idea of 'loose coupling,' in A. C. Kerckhoff (ed.) *Research in the Sociology of Education and Socialization*, Vol II (Greenwich, CT: JAI Press), 261–291.

DAVID, J. L. and PETERSON, S. M. (1984) *Can Schools Improve Themselves?* (San Francisco: Bay Area Research Group).

DAVIES, D. (1980) School administrators and advisory councils: partnership or shotgun marriage, *NASSP Bulletin*, 62–66.

DAVIES, D. (1981) Citizen participation in decision making in the schools, in D. Davies (ed.) *Communities and their Schools* (New York: McGraw-Hill), 84–119.

DAVIES, D. (1987) Parent involvement in the public schools, *Education and Urban Society*, 19(2): 147–163.

DAVIES, D., STANTON, J., CLASBY, M., ZERCHKOV, R. and POWERS, B. (1977) *Sharing the Power: A Report on the Status of School Councils in the 1970s* (Boston: Institute for Responsive Education).

DELACY, J. (1990) *The Bellevue Evaluation Study: Studying the Effects of School Renewal* (Seattle: University of Washington Institute for the Study of Educational Policy).

DELACY, J. (1992) *The Bellevue Evaluation Study* (second report) (Seattle, WA: University of Washington Institute for the Study of Educational Policy).

DRACHLER, N. (1977) Education and politics in large cities, 1950–1970, in J. Scribner (ed.) *The Politics of Education*, Yearbook for the National Society of Education (Chicago: University of Chicago Press), 188–217.

DUKE, D. L., SHOWER, B. K. and IMBER, M. (1981) *Teachers as School Decisionmakers* (Stanford: Stanford University Institute for Research on Educational Finance and Governance).

ELMORE, R. F. (1987) Reform and the culture of authority in schools, *Educational Administration Quarterly*, 23: 60–78.

FAGER, J. (1993) *The 'Rules' still Rule: The Failure of School-based Management/Shared Decision-making in the New York City Public School System* (New York: Parents Coalition for Education in NYC).

FINE, M. (1993) [Ap]parent involvement: reflections on parents, power and urban public schools, *Teachers College Record*, 94: 682–709.

FIRESTONE, W. A. (1989) Educational policy as an ecology of games, *Educational Researcher*, 18(7): 18–24.

FISH, J. (1994, April) Institutionalizing the fiction of site based management, a paper prepared for the annual conference of the American Educational Research Association, New Orleans, LA.

FISHER, A. (1979) Advisory committees – does anybody want their advice?, *Educational Leadership*, 7: 254–255.

FULLAN, M. (1993) Innovation, reform, and restructuring strategies, in G. Cawelti (ed.), *Challenges and Achievements of American Education*, (Alexandria, VA: Association for Supervision and Curriculum Development), 116–133.

GAMSON, W. A. (1968) *Power and Discontent* (Homewood, IL: Dorsey Press).

GAVENTA, J. (1980) *Power and Powerlessness, Quiescence and Rebellion in an Appalachian Valley* (Oxford: Clarendon).

GEARY, L. S. (1992) Review of the literature [on the meaning and measurement of power] and explication of the conceptual framework [for examining political processes], in L. S. Geary, The policymaking process resulting in fiscal policy for special education in Utah, doctoral dissertation, Department of Educational Administration, University of Utah, Salt Lake City, UT, 9–58.

GOLDMAN, P., DUNLAP, D. M. and CONLEY, D. T. (1993) Facilitative power and nonstandardized solutions to school site restructuring, *Educational Administration Quarterly*, 29: 69–92.

GOLDRING, E. B. (1993) Principals, parents and administrative superiors, *Educational Administration Quarterly*, 29: 93–117.

GREENFIELD, W. L. (1991) The micropolitics of leadership in an urban elementary school, in J. Blase (ed.) *The Politics of Life in Schools: Power, Conflict and Cooperation* (Newbury Park, CA: Sage), 161–184.

GRONN, P. C. (1984) 'I have a solution . . .' Administrative power in a school meeting, *Educational Administration Quarterly*, 20(2): 65–92.

GUTTMAN, A. (1989) Democratic theory and the role of teachers in democratic education, in J. Hannaway and R. Crowson (eds) *The Politics of Reforming School Administration* (New York: Falmer Press), 183–200.

HARDY, C. (1987) The contribution of political science to organizational behavior, in J. W. Lorsh (ed.) *Handbook of Organizational Behavior* (Englewood Cliffs, NJ: Prentice-Hall), 96–108.

HARGREAVES, A. (1991) Contrived collegiality: the micropolitics of teacher collaboration, in J. Blase (ed.), *The Politics of Life in Schools: Power, Conflict, and Cooperation* (Newbury Park: Sage), 46–72.

HANSON, E. M. (1981) Organizational control in educational systems: a case study of governance in schools, in S. B. Bacharach (ed.) *Organizational Behavior in Schools and School Districts* (New York: Praeger), 245–276.

HANSON, M. (1991) Alteration of influence relations in school-based management innovations, paper presented to the American Educational Research Association, Chicago, IL.

HECKMAN, P., SCULL, R. and CONLEY, S. (1994) Conflict and consensus: the bitter and sweet in a community-school coalition, unpublished paper, University of Arizona, Tuscon, AZ.

HEMPEL, R. L. (1986) *The Last Little Citadel* (Boston: Houghton Mifflin).

HOYLE, E. (1986) *The Politics of School Management* (London: Hodder and Stoughton).

HUGUENIN, K., ZERCHYKOV, R. and DAVIES, D. (1979) *Narrowing the Gap between Intent and Practice: A Report to Policymakers on Community Organizations and School Decision Making* (Boston: Institute for Responsive Education).

IANNACCONE, L. (1975) *Education Policy Systems: A Study Guide for Educational Administrators* (Fort Lauderdale, FL: Nova).

IANNACCONE, L. (1991) Micropolitics of education: what and why, *Education and Urban Society*, 23: 465–471.

IANNACCONE, L. and LUTZ, F. (1970) *Politics, Power and Policy: The Governance of Local School Districts* (Columbus, OH: Charles E. Merrill).

JENNINGS, R. E. (1980) School advisory councils in America: frustration and failure, in G. Baron (ed.) *The Politics of School Government* (New York: Pergamon Press), 23–51.

JOHNSON, S. M. (1989) School work and its reform, in J. Hannaway and R. Crowson (eds) *The Politics of Reforming School Administration* (New York: Falmer Press), 95–112.

JOHNSON, S. M. (1990) *Teachers at Work: Achieving Success in our Schools* (New York: Basic Books).

KIRP, D. L. (1992) What school choice really means, *Atlantic*, 270(5): 119–132.

LICHTENSTEIN, G., MCLAUGHLIN, M. W. and KNUDSEN, J. (1992) Teacher empowerment and professional knowledge, in A. Lieberman (ed.) *The Changing Contexts of Teaching* (Chicago: University of Chicago Press), 37–58.

LIPSKY, M. (1977) *Street-level Bureaucracy* (New York: Russell Sage).

LITTLE, J. W. (1990) The mentor phenomenon and the social organization of teaching, in C. Cazden (ed.) *Review of Research in Education* (Washington, DC: American Educational Research Association), 297–351.

LORTIE, D. (1969) The balance of control and autonomy in elementary school teaching, in A. Etzioni (ed.) *The Semi-Professions and their Organizations* (New York: Free Press).

LUKES, S. (1974) *Power: A Radical View* (London: Macmillan).

MALEN, B. (1993) 'Professionalizing' teaching by expanding teachers' roles, in S. L. Jacobson and R. Berne (eds) *Reforming Education: The Emerging Systemic Approach* (Thousand Oaks, CA: Corwin Press), 43–65.

MALEN, B. (1994a) Enacting site based management: a political utilities analysis, *Educational Evaluation and Policy Analysis*, 16: 249–267.

MALEN, B. (1994b) Enhancing the information base on Chicago school reform: a commentary on the Consortia's 'governance stream' of research, paper prepared at the request of the Consortia on Chicago School Research (Chicago: University of Chicago Center for School Improvement).

MALEN, B. and HART, A. W. (1987a) Career-ladder reform: a multi-level analysis of initial effects, *Educational Evaluation and Policy Analysis*, 9: 9–23.

MALEN, B. and HART, A. W. (1987b) Shaping career ladder reform: the influence of teachers on the policy-making process, paper prepared for the annual conference of the American Educational Research Association, Washington, DC.

MALEN, B. and OGAWA, R. T. (1988) Professional–patron influence on site-based governance councils: a confounding case study, *Educational Evaluation and Policy Analysis*, 10: 251–279.

MALEN, B., OGAWA, R. T. and KRANZ, J. (1990) What do we know about school-based management? A case study of the literature – a call for research, in W. H. Clune and J. F. Witte (eds) *Choice and Control in American Education*, Vol. II (London: Falmer Press), 289–432.

MANGHAM, I. (1979) *The Politics of Organizational Change* (Westport, CT: Greenwood).

MANN, D. (1974) Political representation and urban school advisory councils, *Teachers College Record*, 75: 270–307.

MANN, D. (1975) Democratic theory and public participation in educational policy decisionmaking, in F. M. Wirt (ed.) *The Polity of the Schools* (Lexington, MA: D. C. Heath), 5–21.

MANN, D. (1977) Participation, representation and control, in J. Scribner (ed.) *The Politics of Education,* Yearbook for the National Society for the Study of Education (Chicago: University of Chicago), 67–93.

MARSHALL, C. (1993) The politics of denial: gender and race issues in administration, in C. Marshall (ed.) *The New Politics of Race and Gender* (New York: Falmer Press), 168–174.

MARSHALL, C. and MITCHELL, B. A. (1991) The assumptive worlds of fledgling administrators, *Education and Urban Society,* 23: 396–415.

MARSHALL, C. and SCRIBNER, J. D. (1991) It's all political, *Education and Urban Society,* 23: 347–355.

MAURIEL, J. J. and LINDQUIST, K. M. (1989) *School-based management doomed to failure?,* paper presented at the annual conference of the American Educational Research Association, San Francisco, CA.

MCLAUGHLIN, M. W. (1987) Lessons from experience: lessons from policy implementation, *Educational Evaluation and Policy Analysis,* 9: 171–178.

MCLEESE, P. (1992) *The process of decentralizing conflict and maintaining stability: site council enactment, implementation, operation and impacts in the Salt Lake City School District, 1970–1985,* doctoral dissertation, Department of Educational Administration, University of Utah, Salt Lake City, UT.

MCNEIL, L. M. (1986) *Contradictions of Control – School Structure and School Knowledge* (New York: Routledge and Kegan Paul).

MCPHERSON, R. B., SAILEY, C. and BAEHR, M. E. (1975) What principals do: a national occupational analysis of the school principalship, *Consortium Currents,* 2: 1–10.

MOLES, O. C. (1987) Who wants parent involvement?, *Education and Urban Society,* 19: 137–145.

MOORE, D. R. (1991, April) Chicago school reform: the nature and origin of basic assumptions, paper prepared for the annual conference of the American Educational Research Association, Chicago, IL.

MOORE, D. R. and DAVENPORT, S. (1989) *The New Improved Sorting Machine* (Chicago: Designs for Change).

MORGAN, G. (1986) *Images of Organizations* (Beverley Hills: Sage).

MUNCEY, D. E. and MCQUILLAN, P. J. (1992) The dangers of assuming a consensus for change: some examples from the coalition of essential schools, in G. A. Hess, Jr (ed.) *Empowering Teachers and Parents: Restructuring through the Eyes of Anthropologists* (New York: Bergin and Garvey), 47–69.

OPOTO, S. (1991) Adolescent peer-conflicts: implications for students and for schools, *Education and Urban Society,* 23: 416–441.

PFEFFER, J. (1981) *Power in Organizations* (Marshfield: Pitman).

POPKEWITZ, T. S. (1979) Schools and the symbolic uses of community participation, in C. A. Grant (ed.) *Community Participation in Education* (Boston: Allyn and Bacon), 202–223.

POWELL, A. G., FARRAR, E. and COHEN, D. K. (1985) *The Shopping Mall High School* (Boston: Houghton Mifflin).

REED, D. B. and MITCHELL, D. E. (1975) The structure of citizen participation: public discussions for public schools, in National Committee for Citizens in Education Commission on Education Governance (eds) *Public Testimony on Public Schools* (Berkeley, CA: McCutchan), 183–217.

ROLLOW, S. G. and BRYK, A. S. (1993a) Democratic politics and school improvement: the potential of Chicago school reform, in C. Marshall (ed.) *The New Politics of Race and Gender* (New York: Falmer Press), 97–106.

ROLLOW, S. G. and BRYK, A. S. (1993b) Grounding a theory of school micro-politics: lessons from Chicago school reform, paper available through the Center for School Improvement, University of Chicago, Chicago, IL.

ROWAN, B. (1990) Commitment and control: alternative strategies for the organizational design of schools, in C. Cazden (ed.) *Review of Research in Education* (Washington, DC: American Educational Research Association), 353–389.

SALISBURY, R. H. (1980) *Citizen Participation in the Public Schools* (Lexington: D. C. Heath).

SARASON, S. B. (1990) *The Predictable Failure of Educational Reform* (San Francisco: Jossey-Bass).

SCHATTSCHNEIDER, E. E. (1960) *The Semi-sovereign People* (New York: Holt, Rinehart and Winston).

SCHWILLE, J., PORTER, A., BELLI, G., FLODEN, R., FREEMAN, D., KNAPPEN, L., KUHS, T. and SCHMIDT, W. (1983) Teachers as policy brokers in the content of elementary school mathematics, in L. S. Shulman and G. Sykes (eds) *Handbook of Teaching and Policy* (New York: Longman), 370–391.

SCHWILLE, J., PORTER, A. and GRANT, M. (1980) Content decision making and the politics of education, *Educational Administration Quarterly,* 16: 21–40.

SCOTT, W. R. (1992) *Organizations Rational, Natural and Open Systems* (Englewood Cliffs: Prentice-Hall).

SEDLAK, M. W., WHEELER, C. W., PULIN, D. C. and CUSICK, P. A. (1986) *Selling Students Short* (New York: Teachers College Press).

SHAVELSON, R. J. (1981) *A Study of Alternatives in American Education: Conclusions and Policy Implications* (Santa Monica: Rand).

SHIELDS, P. and MCLAUGHLIN, M. (1986) *Parent Involvement in Compensatory Education Programs* (Stanford: Center for Educational Research at Stanford).

SMYLIE, M. A. (1992) Teacher participation in school decision making: assessing willingness to participate, *Educational Evaluation and Policy Analysis*, 14(1): 53–67.

SMYLIE, M. A. (1994) Redesigning teachers' work: connections to the classroom, *Review of Research in Education*, 129–178.

SMYLIE, M. A. and TUERMER, U. (1992, February) The politics of involvement versus the politics of confrontation: school reform and labor relations in Hammond, Indiana, report prepared for Project VISION, Claremont Graduate School, Claremont, CA.

SROUFE, G. E. (1977) Politics and evaluation, in J. D. Scribner, (ed.) *The Politics of Education*, the Seventy-sixth Yearbook of the National Society for the Study of Education (Chicago, IL: University of Chicago Press), 287–318.

STEVENSON, K. R. and PELLICER, L. (1992) School-based management in South Carolina: balancing state-directed reform with local decision making, in J. J. Lane and E. G. Epps (eds) *Restructuring the Schools: Problems and Prospects* National Society for the Study of Education (Berkeley, CA: McCutchan), 123–139.

SUMMERFIELD, H. L. (1971) *The Neighborhood-based Politics of Education* (Columbus, OH: Charles E, Merrill).

TOWNSEND, R. G. (1990) Toward a broader micropolitics of schools, *Curriculum Inquiry*, 20: 205–224.

TYACK, D. (1981) Historical perspectives on public education, in D. Davies (ed.) *Communities and their Schools* (New York: McGraw Hill), 11–32.

VITERITTI, J. P. (1986) The urban school district: toward an open system approach to leadership and governance, *Urban Education*, 21(3): 228–253.

WALLER, W. (1932) *The Sociology of Teaching* (New York: Wiley).

WEATHERLEY, R. and LIPSKY, M. (1977) Street-level bureaucrats and institutional innovation: implementing special-education legislation reform, *Harvard Educational Review*, 47: 171–197.

WEISS, C. H. (1993) Shared decisionmaking about what? A comparison of schools with and without teacher participation, *Teachers College Record*, 95(1): 68–92.

WEISS, C. H. and CAMBONE, J. (1993) Principals' roles in shared decisionmaking, revised version of a paper presented at the American Educational Research Association, Atlanta, GA.

WEISS, C. H., CAMBONE, J. and WYETH, A. (1992) Trouble in paradise: teacher conflicts in shared decision-making, *Educational Administration Quarterly*, 28: 350–367.

WHITE, P. A. (1992) Teacher empowerment under 'ideal' school-site autonomy, *Educational Evaluation and Policy Analysis*, 14: 69–82.

WILES, D. K. (1974) Community participation demands and local school response in the urban environment, *Education and Urban Society*, 6(4): 451–468.

WIRT, F. M. and KIRST, M. W. (1975) *The Political Web of American Schools* (Boston: Little, Brown).

WONG, K. K. (1994) Linking governance reform to schooling opportunities for the disadvantaged, *Educational Administration Quarterly*, 30(2): 153–177.

10. Rethinking the public and private spheres: feminist and cultural studies perspectives on the politics of education[1]

Catherine Marshall and Gary L. Anderson

One of the tasks of critical scholarship is to understand how fields of study have been constructed. Yearbooks like this one play a strong legitimization role in that they reinforce what counts as the politics of education. In this chapter we challenge the parameters defining the field of educational politics and the dominance in the field of a single paradigm. These parameters have created a structured silence with regard to the exercise of power and gender, social class, racial, and sexual difference. By legitimizing new paradigms and theories, we hope to broaden the parameters of the field and thus contribute to an exploration of these areas of silence. First, we will briefly review current feminist theories and work in the interdisciplinary field of cultural studies as they relate to educational policy and politics. Then, we will provide an example of how cultural studies and feminist theory combine and explore how our definition of the 'public sphere' determines how we think about educational policy and politics.

New theories, new paradigms

Theories, when used heuristically, are lenses or windows that provide a particular view of social phenomena, opening up vistas not to be seen from other windows/theories. In this way new theoretical perspectives can make visible those aspects of traditional educational phenomena made invisible by previous theoretical frames. New theories can also illuminate previously ignored phenomena, opening up new areas for critical examination. Much as feminist theory in the 1970s opened up the personal as as arena of political struggles, feminist and critical theories have problematized and politicized areas of education that previously, if viewed at all, were viewed as nonproblematic and non-political,

Theories are embedded in paradigms. The dominant paradigm in the politics of education has been characterized by a theory of society that emphasizes a consensus of values, social integration and coherence, and the need for regulation in social affairs (Burrell and Morgan 1979). This paradigm includes theoretical approaches to political phenomena such as political systems theory, bureaucratic theory, exchange theory, social action theory, symbolic interactionism, and theories of negotiated order. In this paradigm, when conflict and lack of consensus occur, they are viewed as temporary aberrations which are 'fixed' through political processes. The nonrational, paradoxical, non-categorized phenomenon or group is seen as a minor disruption.

The last two decades have seen as increasing interest in theories of society that emphasize structural conflict, domination/emancipation, and forms of human agency involving various strategies of appropriation, accommodation and resistance. In this

0268-0939/94 $10·00 © 1994 Taylor & Francis Ltd.

chapter we discuss ways of incorporating feminist, critical and postmodern theories and cultural studies into the study of the politics of education.

Feminist theories and cultural studies

We wish to demonstrate how new theoretical perspectives based on alternative paradigms of social analysis both provide new lenses for understanding the politics of education in traditional arenas and provide new political arenas that have been hidden by the exclusive use of traditional lenses. We will concentrate primarily on the intersection of two theoretical lenses: feminist theories and cultural studies.

Feminist theories have been applied in recent years to the study of pedagogy and curriculum, but the predominantly male field of educational politics has generally not benefited from its theoretical insights. *Cultural studies* are currently enjoying considerable popularity outside the field of education, and provide an interdisciplinary – some would say anti-disciplinary – approach to cultural analysis that incorporates many of the insights of critical, feminist and postmodern theories.

Both feminist theory and cultural studies provide lenses that force one to re-vision the very notion of politics and policy. First, both are concerned with micro-analysis of the more covert ways in which power is exercised. This leads researchers to study those areas not previously defined as relevant political arenas, e.g., popular culture, informal aspects of organizational and social life, and personal experience.

Second, while politics of education researchers acknowledge that social institutions represent negotiated orders (Bazerman and Lewicki 1983), feminist theories and cultural studies insist on viewing social negotiation as occurring within unequal and often hidden power relations based on class, race, gender, and sexual orientation.

Third, feminism and cultural studies challenge the separation between theorizing and acting. According to Grossberg (1994) 'cultural studies is both an intellectual and a political tradition . . . where "culture" is simultaneously the ground on which analysis proceeds, the object of study, and the site of political critique and intervention' (p. 5). The same could be said of feminist theory with regard to women's personal and social experiences.

We will demonstrate these points through illustrative vignettes, through brief overviews of cultural studies and feminist theory, and by identifying the intersect between these theorists and policy analysis. We will then demonstrate how these theories expand our definitions of politics and our notions of the public and private spheres.

Feminist theories: challenging old definitions, opening new arenas

The federal education official (female!) declared: 'The problem of sex equity in schooling has been handled by Title IX long ago . . . . I see no need for further study and legislative action' and the activists who sponsored the AAUW report watched helplessly as their carefully constructed study, aimed at generating new legislation with teeth, was simply tossed aside as the federal official's words fit with the master narrative that framed policy issue identification.

As the school board meeting agenda continued toward midnight Estelle (Assistant Superintendent) and George (Superintendent) were on a roll, getting their plans approved with dwindling resistance and a dwindling audience. In the back of her mind, Estelle was preparing for the awkward end-of-meeting debriefing, alone with George, with both of them feeling victorious and close. Intuitively she knew that part of her job was to prevent the good feelings from escalating into sexual feelings; if she wanted to keep her job she had to make sure that she, George, George's wife, community members – that nobody would view their relationship as threatening or suspect.

These vignettes illustrate areas in which feminist theory has the potential to inform the politics of education. The first vignette sets the stage for discussing some of the differences that exist *within* feminist theory and that will be reviewed in the section. The second vignette illustrates those informal arenas of politics that are ignored by traditional scholarship, but which are recognized by women as a crucial aspect of educational politics.

Writing against the grain: a diversity of perspectives

Feminist theory 'brings to consciousness facets of our experience as women that have . . . contradicted predominant theoretical accounts of human life' (Keohane *et al.* 1981: vii). Feminist education scholarship is marginalized, flourishing only when it focuses on classrooms and teachers, on children and women. Therefore, feminist scholarship offers a challenge to scholars of educational politics to call attention to power and political dynamics that have, heretofore, been occluded by embedded male domination of institutions and thought. Politics of education scholars seldom focus on gender; a feminist politics of education exists only outside the mainstream (malestream) of politics and policy analysis. Feminist theories point to new arenas of political contestation and provide possible new lenses and tools for reconfiguring current masculinist models, ultimately benefiting both men and women.

This section provides a brief overview of three current perspectives within feminist thought: liberal feminism, difference feminism, and power and politics feminism. This overview provides the grounding for a challenging new field: feminist critical policy analysis in education, or feminist politics of education.

As with any theoretical developments, feminism has a history of debates and critiques (Eisenstein 1993, Tong 1989, Weiler 1988). One strand of feminist theories and research emphasizes the barriers to females' access and choice. This *Liberal Feminism* strand has recognized how differential socialization and opportunity have limited women. It has generated liberal policies – laws that assume that simply eliminating barriers and placing women in positions will change institutional and cultural values, a politically naive stance (Ferguson 1984, Marshall 1993). Liberal feminism is the basis for policies such as affirmative action, the Equal Rights Amendment, and comparable worth. While working for equity has obvious merit, what is missing is the realization that the people with power in political, institutional, and professional cultures that created sexist and differential access are being relied on to create new power and access processes and to willingly and thoughtfully give up their power and privilege.

Difference Feminism is a second strand emphasizing and appreciating women's perspectives and differences. Tong calls this psychoanalytical feminism; Eisenstein calls it the 'women's studies' strand. It differs from traditional sex roles socialization theory in that Difference clearly posits a need to value women's ways. By demonstrating that women have different socialization, different orientations to moral decisionmaking, ways of knowing, and ethics (Belenky *et al.* 1986, Chodorow 1978, Gilligan 1982, Noddings 1988), this strand names and values women's subjective experience. This strand makes an important contribution; it validates women's studies; it provides a challenging critique to every field whose theory and research base has excluded women, from medicine to political science to literature, to history and all sciences. However, this strand goes nowhere until it presses on the political nature of the process by which men's ways, socialization experiences, and needs have become the mainstream legitimate ones, embedded in all institutions, especially in politics. As important, this Difference strand tends to homogenize women (and men), (the 'essentialist' critique) as if there were one women's experience or one feminism, ignoring class, race, political and sexual orientation, and other within-women differences.

Another strand emphasizes power. *Power and Politics Feminism* includes the cultural reproduction theorists, the radical critique, Marxist, socialist, existentialist, and postmodern strands of feminism. All identify the institutional, economic purposes and the political and cultural processes which create and maintain exclusion of females. Socialist feminist thought emphasizes how the power of capitalism and patriarchy combine to oppress women. Existentialist feminism points to the myths and stereotypes created to maintain women in the place of 'Other' – with men in control of myth-creation and maintenance and men deciding women's feminine identity. Postmodern feminism also recognizes power, assuming that all structures are social constructions, created for some political purpose, often hurting women, and targets for deconstruction. For example, conventions for behavior and discourse and norms for leaders and for professionals, which are for demonstrating competence, are artifices; when they exclude women, they become part of sexual politics.

This Power and Politics theoretical strand recognizes that simply gaining power in the context of existing power structures must be rejected. 'Much feminist thinking has, therefore, gone into seeking ways to restructure power relations' (Boneparth 1982: 14). MacKinnon (1982) cautions against simplistically seeking representation in political spheres and power to create the world from one's own point of view, pointing out that these are the political strategies of men, combining legitimization with force. Instead, MacKinnon argues, 'feminism revolutionizes politics' (p. 3).

Thus, a range of feminisms emphasize the subtle politics of women's oppression. The Power and Politics strand holds the most promise for combining the study of politics and gender because it identifies the ways in which political systems act as societally constructed institutions which reproduce gendered power relations.

With these strands of feminist theories as the undergirding substance, politics of education scholars now have the opportunity to conduct a new kind of policy analysis. Feminist Critical Policy Analysis is new and rare; it begins with the assumption that gender inequity results from purposeful (if subconscious) choices to serve some in-group's ideology and purpose. It is research concerned with identifying how the political agenda benefiting males is embedded in school structures and practices. Feminist critical policy analysis is research that conducts analyses *for* women while focusing on policy and politics, and it asks about every policy or political action, 'how does this affect females?,' an often neglected question. Reforms of schooling have 'failed to challenge the "male-as-norm" conceptions of educational purpose, of students, of teachers, of curricula, of pedagogy, and of the profession of education' (Leach and Davies 1990: 322). Feminist curriculum theorists, education philosophers, and researchers of teaching careers and classroom dynamics have made important contributions to the field of education, but few scholars in politics and policy have been engaged with gender issues and feminist research. Feminist critical policy analysis will end that silence.

In the following section we describe another new theoretical perspective, cultural studies, and the ways it refines politics as 'cultural politics' and draws on feminist and other critical traditions.

Cultural studies and cultural politics

Patrice is a young Black female, in eleventh grade. She says nothing all day in school. She sits perfectly mute. No need to coerce her into silence. She often wears her coat in class. Sometimes she lays her head on her desk. She never disrupts. Never disobeys. Never speaks. And is never identified as a problem. Is she the student who couldn't develop two voices and so silenced both? Is she so filled with anger, she fears to speak? Or so filled with depression she knows not what to say? (Fine 1991: 50–51)

After the new Dean announced the list of candidates for the four positions of associate dean to the college faculty, a Latina faculty member raised her hand. In a halting voice, she asked if any consideration had been given to including ethnic minority candidates, pointing out the large percentage of ethnic minorities in the southwestern state in which the college was located. There was silence around the room. The Dean indicated that she had approached a couple of minority faculty about the positions and that they had shown no interest in them. This seemed to satisfy the faculty that the real problem was not racism, but rather a lack of interest on the part of minority faculty in being part of the Dean's administrative team, and the Dean moved on to the next agenda item.

Scenes like these take place in educational institutions every day. The apparent passivity of Patrice and that of the Hispanic faculty who showed no interest in the associate dean positions is assumed to be mere passivity rather than passive resistance. To view the behavior as passive resistance would raise questions about what was being resisted and why. Educational institutions are rife with countless forms of active and covert resistance. The Latina faculty member described above continues actively to challenge injustice in the system, but her active forms of resistance have earned her a series of labels that serve to marginalize her and discredit her views. Patrice, the eleventh grader, and the other minority faculty have simply chosen not to play the game, opting instead for the more covert resistance that their silence and apparent passivity symbolize.

To the extent that both active and passive forms of institutional resistance can be turned back onto the individual, institutional legitimacy is maintained. The individual is labeled, marginalized, pathologized in order to protect the legitimacy of the institution. A cultural studies approach to the politics of education makes the claim that a central role of educational administrators and policy makers is to maintain the legitimacy of the status quo by playing a gatekeeper role over the policy agenda and by 'managing' conflict and silencing resistance.

Cultural studies: reframing old questions and constructing new ones

The above examples raise questions that are the foci of cultural studies: How are some events defined as political and others as nonpolitical? How is politics mediated by culture? Whose interests are served by differing definitions of the cultural and political? Within a critical framework, culture is no longer merely a way of life, but must be understood as a form of production that takes place within asymmetrical power relations. Quantz (1992) argues that we cannot talk about culture without talking about some conception of cultural politics. According to Quantz, 'culture is not so much the area of social life where people share understandings as that area of social life where people struggle over understandings' (p. 487). Quantz cites Keesing (1987), who, playing on Geertz's famous definition of culture, remainds us that 'cultures are webs of mystification as well as signification' (p. 487).

The Cultural Studies tradition has promoted two principal shifts in the locus of analysis of educational politics and policy. The first shift involves the relationship between the macropolicy arena and its local or 'street-level' impact. Traditional models have viewed the policy filtering process as flawed but capable of being rationalized through better implementation models. Cultural studies suggest that micropolitics at the local level involves complex forms of cultural and political resistance, accommodation, and compliance rooted in the informed intentionality of social actors. The second is a shift from viewing education primarily as schooling, to an analysis of the impact on social actors of popular culture, the media, and other institutions that 'educate' (see Roman *et al.* 1988, Giroux and McLaren 1994).

Both of these shifts are, in part, due to the influence of the Centre for Contemporary Cultural Studies at Birmingham, England, in which cultural and policy analyses were done simultaneously. This early work focused on reconceptualizing the role of the state in educational policy. While structural Marxists tended to view schools as part of a network of 'state ideological apparatuses' (Althusser 1971) that imposed ideology from the top down, Marxism lacked a cultural theory that aided in understanding the complexity of the cultural politics that took place within these social institutions on an everyday basis. Thus, Willis (1977) and McRobbie (1980, 1991) submitted social reproduction theory to cultural analysis and found that 'school failure' was a complex social construction involving a dynamic and interactive relationship between youth subcultures and the dominant culture represented by the school. North American studies, inspired by educational ethnography and critical and feminist theory, have provided numerous cultural analyses of school subcultures and educational policy (see Fine 1991, Wexler *et al.* 1992). They have called into question dominant notions like 'drop outs', 'at-risk' students, and 'conflict management' and have brought issues like institutional silencing and marginalization, identity negotiation, resistance, and popular culture to center stage. Should the field of educational politics and policy continue to ignore this important work, then it will become increasingly irrelevant to the present crisis of our public school system.

Applying feminist and cultural studies perspectives: defining politics and the public sphere

One of the areas of analysis which has become increasingly important for educational politics in recent years is the definition of the public sphere. Feminist theory and cultural studies have led the way in analyzing the complex relationships between the public and the private. This section discusses different ways in which feminism and cultural studies are currently reconceptualizing the public and the private spheres.

The public sphere as male and the private sphere as female

While the relation of public and private spheres to issues of gender is a common premise of gender studies, it is seldom acknowledged in the field of educational policy and politics. The separation of the public and private spheres and the corresponding separation of the male and female worlds have their origins in the rise of capitalism. Although patriarchy predates capitalism, it is with the advent of capitalism that the separation of the public and the private spheres intensifies. Smith (1987) provides the following account of this process:

> The forms that the power of men over women have taken are located in relations coinciding with relations organizing the rule of dominant classes. Historically the organization of these relations and their dynamic expansion are intimately linked to the dynamic progress of capital. Capitalism creates a wholly new terrain of social relations external to the local terrain and the particularities of personally mediated economic and social relations. It creates an extralocal medium of action constituted by a market process in which a multiplicity of anonymous buyers and sellers interrelate and by an expanding arena of political activity. These extralocal, impersonal, universalized forms of action became the exclusive terrain of men, while women became correspondingly confined

to a reduced local sphere of action organized by particularistic relationships . . . . The differentiation of public and private which we have come to take for granted is structured in this progressively massive shift. Formerly common to both women and men, the domestic became a discrete and lesser sphere confining and confined to women and on which the domain arrogated by men has continually encroached. (p. 5)

For minority women, this relegation to the private sphere often meant that their work – domestic work – was done in dominant culture homes. The differences in status which have resulted from this division of the public and private spheres have carried over into those areas of public life in which women predominate. These differences have negatively affected the status and economic rewards of feminized professions such as teaching, social work, and nursing and reduced parenting to an unrecognized and unvalued part of 'women's work.' Meanwhile, as women enter the public sphere, primarily by sheer economic necessity, the domestic sphere remains the responsibility of women. The now commonly acknowledged phenomenon of the 'double shift' for women who must work both inside and outside the home is thus produced. These issues and many others related to this public/private/gender dynamic, while heavily theorized in other disciplines, remain almost untouched in the areas of educational policy and politics. The politics of curriculum as well as of teacher and leadership training tends to accept as unproblematic the built-in assumption that the work of women is un- or under-valued.

The private in the public and the public in the private

Traditional definitions of politics include: 'who gets what when and how' (Harold Lasswell), 'the authoritative allocation of values' (David Easton), and 'turning personal troubles into public issues' (C. Wright Mills). The last definition, with cultural and feminist perspectives added, raises questions about which personal troubles stay personal, and which (or whose) troubles get public attention and resources in political arenas. The Greek polis designated the affairs of male citizens to be the public sphere; all else was the realm for women and slaves. Nowadays, politicians (mostly male) negotiate over the public nature of issues like abortion, child care, maternity leave, adolescent pregnancy, and sexual harassment.

Schmukler (1992) describes a 'public–private border zone' in which, through constant interaction between social actors and public institutions, private needs are turned into demands on the public world. Constant negotiation takes place in this zone as formerly domestic concerns like those mentioned above become the concern of the public sphere (Wexler 1982). Furthermore, as women increasingly enter the public sphere, they bring with them many of the ways of knowing that have the potential to change the nature of organizational politics. This notion from difference feminism, however, requires a power and politics feminism to understand what happens as characteristics from the private sphere are taken into the public one. Ferguson (1984) has demonstrated how the potential for femaleness to democratize organization is thwarted by hierarchical power relations:

As long as women are subordinate to men, the virtues of female experience will be turned to the requirements of surviving subordination: the capacity to listen, to empathize, to hear and appreciate the voice of the other, and so forth, will be used as strategies for successful impression management (p. 186)

The cultural and feminist approach to the micropolitics of organizational life contributes to an understanding of the distorting effects of power and the ways that power is exercised invisibly. However, these insights are seldom evident in the work of those who seek to democratize the public sphere at the institutional level. Educational journals are full of accounts by scholars and practitioners of how to 'empower teachers' and 'involve students, parents and the community.' Most of this literature lacks even a rudimentary understanding of the many subtle ways that power is exercised in the public sphere to stifle democracy (see Anderson 1990, Dunlap and Goldman 1991, Malen and Ogawa 1988, and Robinson 1994 for exceptions).

Not only does the feminist and cultural studies approach illuminate private issues within the public sphere such as the sexual politics described in the above vignette about Estelle and George, it also shows how the public sphere has insidiously invaded the private, disciplining and administering even our most intimate thoughts and acts. Bookchin (1978) describes how the increasing bureaucratization and administration of the private sphere replace traditional social forms, undercutting the kind of sociality and solidarity necessary for social opposition:

> Under capitalism, today, bureaucratic institutions are not merely systems of social control; they are literally institutional substitutes for social form . . . . However much market society may advance productive forces, it takes its historical revenge not only in the rationalization it inflicts on society, but the destruction it inflicts on the highly articulated social relations that once provided the springboard for a viable social opposition. (p. 18)

Community is replaced by clusters of autonomous individuals, each with a personal set of individual concerns. Individuals become de-politicized as bureaucratic and market rationality make the notion of community and solidarity, at best, an inconvenience and, at worst, a bad risk. Ferguson (1984) has perhaps most eloquently described the dangers of this expansion of the public sphere:

> Freedom in bureaucratic capitalism refers to an arena of privacy surrounding each individual, protected from encroachments by other individuals; community, to the extent the word can be used at all, is a secondary arrangement among already autonomous beings. (pp. 196–197)

Constrained values and structures in the public sphere

Lourde's (1984) provocative statement, 'You cannot dismantle the master's house using the master's tools' (p. 112), introduces an important critique of accepted public sphere structures. The structures that, over centuries, have functioned to channel public policy debates and policy alternatives in ways that exclude large groups, even in democratic societies, are not the best structures for dismantling the ways privileges are currently structured. In their analysis of gender issues in leadership, Ortiz and Marshall (1988) said:

> Perhaps those already in charge would not permit themselves and strategically located policymakers to consider new forms and structures that would challenge existing patterns of control and distribution of power. (p. 136)

Stromquist's (1993) review of sex equity policy implementation confirms this explanation: with little effort at monitoring, training, or enforcement, and with ample attention to protecting dominant interests from any ill effects of sex equity, gender equity is still problematic after 20 years of policy nonimplementation.

Thus, traditional policy activity, policy analysis and policy studies will not sufficiently capture the issues of the disempowered. The master's tools do not work for them; they may create alternative tools, but traditional policy analysts may not recognize them as policy initiatives.

Similarly, the accepted values in the public sphere may not capture those envisioned by women. Traditionally, education policy analysts document the ways in which public domain values clash. For example, Marshall et al. (1989) studied how equity, choice, quality, and efficiency values directed policy choices in state legislatures. However, from 'difference feminism' new ways of viewing values, ethics, decisionmaking, and leadership are emerging. The traditional boundaries of the public sphere (and scholars of policy and politics) have not incorporated these new perspectives.

Women's experiences have led to women having different ways of knowing and valuing. Gilligan (1982) demonstrates that women's moral decisionmaking is more attuned than men's to relationships and caring; Noddings (1992) proposes an ethic of caring as the basis for behaviors and structures for schooling; Regan (1990) critiques the masculinist, corporate-styled hierarchical model for school management and displays a feminist leadership that creates connections and collaboration.

Traditional policy tools and traditional analyses of values work only with a constrained view of the public sphere. Politics of education scholars must work to expand that view.

The expanding marketplace: the privatization of the public sphere

In his speech announcing the creation of the Task Force on Educational Excellence, the Governor said, 'I am so pleased that [the CEOs of two of the major industries] have agreed to Co-Chair this vitally important effort. To formulate standards of excellence for schools without direction from our business leaders would be like spitting in the wind.'

The economic literature in education identifies a need to use market principles to create greater accountability (e.g., school choice, vouchers, etc.) and business as a source of extra resources in a time of economic cutbacks (e.g., school–business partnerships). It has also been framed by neo-liberal economists as the privatization of the public sector of our economy. Fueled by the analyses of pro-voucher researchers (Chubb and Moe 1986) and the political muscle of corporations like Whittle Communications, many policy makers seem willing to privatize major segments of our public school system. Scholars like Apple (1993) have further pointed out that, in a world of greater competition and shrinking markets, the public schools represent a major new market to be opened up and developed.

Historians have documented how the management systems, bureaucracies, and leadership models from corporate America have been embedded in schools, perpetuating class, race, and gender differentiations (Callahan 1962, Katz 1971, Tyack 1974). Thus, the basic structure of schooling makes it amenable to further privatization. The deep-seatedness of market assumptions makes it difficult to end the silences and raises questions about teacher-proof curricula, policies that separate high achievers from low achievers, or policies that increase choice, resources, and access only for advantaged groups. Confusions about distinctions between the public and private economic spheres have become commonplace. This easy and unexamined transfer of economic and market principles to the public discussion of profound normative issues with the logic of efficiency and market forces:

The rhetoric of economic privacy . . . seeks to exclude some issues and interests from public debate by economizing them; the issues in question here are cast as impersonal market imperatives or as 'private' ownership prerogatives or as technical problems for managers and planners, all in contradistinction to public, political matters. (Fraser 1994: 90)

Unless scholars in the politics of education can deepen their analyses of how these political and economic effects will impact on our conception of a public sphere, we will soon find ourselves with a public sphere that is colonized by corporate interests, while we hold on to trivial definitions of democracy and participation that merely serve to legitimize those interests.

Theorizing the public sphere: a feminist encounter with critical theory

Feminism and cultural studies can reframe current debates within the politics of education at both macro and micro levels, providing interdisciplinary dialogue and critique among feminist, critical, and postmodern theories.

Critical theorists relentlessly pursue analyses aimed at identifying and eradicating institutional practices that prevent democratization. For example, Habermas is concerned with creating a public sphere in which communication can proceed without distortion. He describes the public sphere as:

> . . . a theater in modern societies in which political participation is enacted through the medium of talk. It is the space in which citizens deliberate about their common affairs, hence, an institutionalized arena of discursive interaction. This arena is conceptually distinct from the state; it is the site for the production and circulation of discourses that can in principle be critical of the state. The public sphere in Habermas's sense is also conceptually distinct from the official economy; it is not an arena of market relations but rather one of discursive relations . . . . (Fraser 1994: citing Habermas 1989)

In Habermas's ideal, public sphere discussion would be accessible to all; inequalities of status would be bracketed; no merely private interests would be admissible. Public deliberation would have outcomes which represent the common good. The power usually exercised by dominant interests would no longer distort public debate.

While Habermas has gained uncritical acceptance among critical theorists in education, Fraser's (1994) work provides an example of a feminist cultural studies analysis of educational politics. She criticizes Habermas and calls for a reconceptualization of the public sphere.

According to Fraser (1994), Habermas's critical theory of the public sphere makes conceptual errors common to mainstream modernist narratives of politics. First, it assumes, as does much politics of education literature, a single public sphere in which all can participate. Fraser points out that multiple 'hidden' public spheres have always existed, including women's voluntary associations, working-class organizations, and others, and that a plurality of public spheres is preferable to Habermas's notion of a single, common one. She cites the existence of 'counterpolitics' composed of subordinated social groups, such as women, workers, peoples of color, and gays and lesbians. These counter-publics

> . . . invent and circulate counterdiscourses, which in turn permit them to formulate oppositional interpretations of their identities, interests, and needs. Perhaps the most striking example is the late-twentieth century US feminist subaltern counterpublic with its variegated array of journals, bookstores, publishing companies, film and video distribution networks, lecture series, research centers, academic programs, conferences, conventions, festivals and local meeting places. In this public sphere, feminist women have invented new terms for describing social reality, including 'sexism: the double shift,' 'sexual harassment,' and 'marital, date, and acquaintance rape.' (Fraser 1994: 84)

Fraser further argues that Habermas's claim of openness to participation vastly under-estimates the subtle ways in which individuals and groups are excluded from participation in the public sphere. Fraser points out that discursive interaction within the largely male-dominated bourgeois public sphere has always been governed by protocols of style and decorum that were themselves markers of status inequality. Recent politics of education research confirms that issues of cultural capital operate in shared governance structures in schools (Malen and Ogawa 1988), but Fraser takes the notion a step further. She suggests that public spheres set up for open deliberation can serve as a mask for domination. She cites Mansbridge (1990) who asserts that:

> . . . the transformation of 'I' into 'we' brought about through political deliberation can easily mask subtle forms of control. Even the language people use as they reason together usually favors one way of seeing things and

discourages others. Subordinate groups sometimes cannot find the right voice or words to express their thoughts, and when they do so, they discover they are not heard, [They] are silenced, encouraged to keep their wants inchoate, and heard to say 'yes' when what they have said is 'no.' (p. 127)

The relevance to participatory decisionmaking in schools and school districts should be clear: as we attempt to set up shared governance structures by creating public spheres for deliberation, we seldom acknowledge competing counterpublics that exist among *and* within school constituencies, much less the subtle ways that domination is exercised within the formal public spheres that are created for democratic deliberation, i.e., school boards, school restructuring councils, curriculum committees, Individualized Education Plan (IEP) meetings, staff meetings, schoolwide assemblies, etc.

Moreover, in economically and socially stratified societies culture and political economy have a synergistic relationship. Not only do unequally empowered social groups tend to develop unequally valued cultural styles, but they have less access to the privately owned, for-profit media that constitute the material support for the circulation of views. Thus, 'political economy enforces structurally what culture accomplishes informally' (Fraser 1994: 82).

Fraser (1994) provides educational politics with several critical insights. First, she challenges critical theorists' conception of the public sphere as an unrealized Utopian ideal, and instead views it as a 'masculinist ideological notion that functioned to legitimate an emergent form of class rule' (p. 79). Second, she conceives of the public sphere as historically constituted by conflict among multiple counterparts. Finally, and perhaps most important for educational politics, she suggests that structures for open deliberation in public spheres can become mechanisms that mask domination.

Theorizing boundaries between public and private

The politics of education has failed to theorize adequately the boundaries between the public and the private. Feminists, for example, have pointed out that the ways of defining issues like domestic violence as part of the domestic private sphere have kept these issues from being dealt with as widespread systemic features of male-dominated societies. Feminists have noted how, in social welfare public policy, social workers are free to invade the private sphere of those dependent upon assistance such as AFDC and food stamps (Connell 1987, Ferguson 1984, Franzway *et al.* 1989, Hill-Collins 1990). In education, feminist scholars have had to fight against assumptions that girls' vocational choices, adolescent pregnancies, and women's unequal access to educational leadership are matters of private choice, not issues for the public policy arena. Politics of education scholars need to study the negotiations which determine whether an issue is deemed private or public.

Students in schools construct their identities within the borderlands between the public and private. Recent studies using narrative and ethnographic research methods are exploring the cultural politics of those who work in and are clients of educational institutions. Anderson and Herr (1993), Fordham (1988), Wexler *et al.* (1992) and many others are attempting to link the identity politics of being a student to institutional structures and practices. Similarly, feminist scholars are exploring issues of silencing and resistance among adolescent girls (Fine 1992, Gilligan and Mikel Brown 1992, Gilligan *et al.* 1990).

The social construction of the public sphere

Perhaps the most important statement from feminist cultural studies is the recognition that the accepted notions about the appropriate public sphere are socially constructed notions. Furthermore, they are constructed to provide advantages to those who had the power to construct them, usually white males. That people accept the dominant notions of the public sphere enhances its power. Other ways, alternative voices, and differently framed priorities are delegitimized as personal problems or fringe-group interests.

Fraser's and other cultural studies approaches to politics and policy (see Bennett 1992) make important linkages to other feminist and cultural accounts of the cultural politics of identity, silencing, and resistance. Until this research begins to appear in reference lists of scholars of educational politics, a rigid and narrow definition of the political will continue to restrict our research to a single paradigm.

Conclusion

In this chapter, we have attempted to suggest possible linkages between research in the field of educational politics and current work in feminist theory and cultural studies, particularly that which reconceptualizes received notions of the public sphere. Our effort has been to legitimize a broadening of current parameters of educational politics to include alternative theoretical perspectives and phenomena that have previously not been viewed as political.

This brings us back to Estelle, the assistant superintendent, who must manage the sexual politics of central office; Patrice, the high school student who sits in stony silence because she's lost her voice, and the outspoken Latina faculty member who is marginalized because she has retained hers. Why has the field of educational politics overlooked this level of cultural politics?

The politics of education as a field of study has historically been closely allied with the field of educational administration, which has tended to use a management lens to view political phenomena. Because of this connection to educational administration, leadership, and policy analysis, it may seem 'unhelpful' to analyze politics from the perspective of the participants in educational institutions. Taking the side of institutional resisters – both active and passive – like those cited above may even seem heretical in a field that seems more interested in impression management and the politics of legitimization and denial than in confronting the causes of persistent social inequities.

Feminist and cultural studies approaches to the politics of education represent a shift away from the management oriented, top-down view of social phenomena, allowing us to understand by 'standing under' our units of analysis. Viewing the public and private spheres in which politics is constructed from this understanding perspective is not unlike viewing America's 'discovery' from the viewpoint of those who were its victims rather than its beneficiaries.

A paradigm shift in the politics of education field will require us to ask new questions and see previously invisible social phenomena. While this kind of shift may make little sense in a 'power over' world, it makes a great deal of sense in the 'power with' world that we hope to create in our educational institutions.

Note

1. The authors wish to express thanks to Edith Rusch and Jean Patterson for their substantive suggestions on this chapter.

References

ALTHUSSER, L. (1971) Ideology and the ideological state apparatuses, in L. Althusser, *Lenin and Philosophy and other Essays* (New York: Monthly Review Press).

ANDERSON, G. (1990) Toward a critical constructivist approach to educational administration, *Educational Administration Quarterly*, 26: 38–59.

ANDERSON, G, and HERR, G. (1993) The micro-politics of student voices: moving from delivery of bodies to diversity of voices in schools, in C. Marshall (ed.) *The New Politics of Race and Gender* (New York: Falmer Press), 58–68.

APPLE, M. (1993) *Official Knowledge: Democratic Education in a Conservative Age* (New York: Routledge).

BAZERMAN, M. H. and LWEICKI, R. J. (eds) (1983) *Negotiating in Organizations* (Beverly Hills, CA: Sage).

BELENKY, M. F., CLINCHY, B., GOLDBERGER, N. R. and TARULE, J. M. (1986) *Women's Ways of Knowing: the Development of Self, Voice, and Mind* (New York: Basic Books).

BENNETT, T. (1992) Putting policy into cultural studies, in L. Grossberg, C. Nelson and P. Treichler (eds) *Cultural Studies* (New York: Routledge), 23–37.

BONEPARTH, E. (1982) A framework for policy analysis, in E. Boneparth (ed.) *Women Power and Policy* (New York: Bergin and Garvey).

BOOKCHIN, M. (1978) Beyond neo-Marxism, **Telos**, 36: 5–28.

BURRELL, G. and MORGAN, G. (1979) *Sociological Paradigms and Organizational Analysis* (Portsmouth, NH: Heinemann).

CALLAHAN, R. (1962) *Education and the Cult of Efficiency* (Chicago: University of Chicago Press).

CHODOROW, N. (1978) *The Reproduction of Mothering: Psychoanalysis and the Socialization of Gender* (Berkeley: University of California Press).

CHUBB, J. E. and MOE, T. M. (1986) No school is an island: politics, markets, and education, *Brookings Review*, 4: 11–27.

CONNELL, R. W. (1987) *Gender and Power* (Oxford: Polity Press).

DUNLAP, D. and GOLDMAN, P. (1991) Rethinking power in schools, *Educational Administration Quarterly*, 27(1): 5–29.

EISENSTEIN, H. (1993) A telling tale from the field, in J. Blackmore and J. Kenway (eds) *Gender Matters in Educational Administration and Policy* (London: Falmer), 1–8.

FERGUSON, K. E. (1984) *The Feminist Case against Bureaucracy* (Philadelphia: Temple University Press).

FINE, M. (1991) *Framing Dropouts: Notes on the Politics of an Urban Public High School* (Albany: SUNY Press).

FINE, M. (1992) *Disruptive Voices: The Possibilities of Feminist Research* (Ann Arbor: University of Michigan Press).

FORDHAM, S. (1988) Racelessness as a factor in Black students' school success: pragmatic strategy or Pyrrhic victory?, *Harvard Educational Review*, 58: 54–84.

FRANZWAY, S., COURT, D. and CONNELL, R. W. (1989) *Staking a Claim: Feminism, Bureaucracy and the State* (Cambridge: Harvard University Press).

FRASER, N. (1994) Rethinking the public sphere: a contribution to the critique of actually existing democracy, in H. A. Giroux and P. McLaren (eds) *Between Borders: Pedagogy and the Politics of Cultural Studies* (New York: Routledge), 74–98.

GIILIGAN, C. (1982) *In a Different Voice* (Cambridge: Harvard University Press).

GILLIGAN, C., LYONS, N. and HANMER, J. (eds) (1990) *Making Connections: The Relational Worlds of Adolescent Girls at Emma Willard School* (Cambridge: Harvard University Press).

GILLIGAN, C. and MIKEL BROWN, L. (1992) *Meeting at the Crossroads: Women's Psychology and Girls' Development* (Cambridge: Harvard University Press).

GIROUX, H. and MCLAREN, P. (eds) (1994) *Between Borders: Pedagogy and the Politics of Cultural Studies* (New York: Routledge).

GROSSBERG, L. (1994) Introduction: bringin' it all back home – pedagogy and cultural studies, in H. A. Giroux and P. McLaren (eds) *Between Borders: Pedagogy and the Politics of Cultural Studies* (New York: Routledge), 74–98.

HABERMAS, J. (1989) *The Structural Transformation of the Public Sphere: An Inquiry into a Category of Bourgeois Society* (Cambridge, MA: MIT Press).

HILL-COLLINS, P. (1990) *Black Feminist Thought* (New York: Routledge).

KATZ, M. (1971) *Class, Bureaucracy, and Schools: The Illusion of Educational Change in America* (New York: Praeger).

KEESING, R. (1987) Anthropology as interpretive quest. *Current Anthropology*, 28(2): 161–176.

KEOHANE, N., ROSALDO, M. Z. and GELPI, B. C. (eds) (1981) *Feminist Theory: A Critique of Ideology* (Chicago: University of Chicago Press).

LEACH, M. and DAVIES, B. (1990) Crossing boundaries: educational thought and gender equity, *Educational theory*, 40(3): 321–332.

LOURDE, A. (1984) *Sister Outsider* (Freedom, GA: Crossing Press).

MACKINNON, C. A. (1982) Feminism, Marxism, method, and the state, in Keohane, N., Rosaldo, M. Z. and Gelpi, B. C. (eds) *Feminist Theory: A Critique of Ideology* (Chicago: University of Chicago Press).

MALEN, B. and OGAWA, R. (1988) Professional–patron influence site-based governance councils: a confounding case study, *Educational Evaluation and Policy Analysis*, 10(4): 215–270.

MANSBRIDGE, J. (1990) Feminism and democracy, *American Prospect*, 1: 125–139.

MARSHALL, C. (1993) The new politics of race and gender, in C. Marshall (ed.) *The New Politics of Race and Gender* (London: Falmer Press), 1–8.

MARSHALL, C., MITCHELL, D. and WIRT, F. (1989) *Culture and Educational Policy in the American States* (London: Falmer Press).

MCNEIL, L. (1986) *Contradictions of Control* (New York: Routledge).

MCROBBIE, A. (1980) Working class girls and the culture of femininity, in Women Study Group, Center for Contemporary Cultural Studies, *Women take Issue* (London: Hutchinson), 96–108.

MCROBBIE, A. (1991) *Feminism and Youth Culture* (Boston: Unwin Hyman).

NODDINGS, N. (1988) An ethic of caring and its implications for American education, *American Journal of Education*, 96(2): 215–230.

NODDINGS, N. (1992) *The Challange to Care in Schools: an Alternative Approach to Education* (New York: Teachers College Press).

ORTIZ, F. I. and MARSHALL, C. (1988) Women in educational administration, in N. Boyan (ed.) Handbook of Research in Educational Administration (New York: Longman), 123–142.

QUANTZ, R. (1992) On critical ethnography (with some postmodern considerations) in M. LeCompte, W. Millroy and J. Preissle (eds) *The Handbook of Qualitative Research in Education* (San Diego: Academic Press), 447–505.

REGAN, H. B. (1990) Not for women only: School administration as a feminist activity. *Teachers College Record*, 9(4): 565–577.

ROBINSON, V. (1994) The practical promise of critical research in educational administration, *Educational Administration Quarterly*, 30(1): 56–76.

ROMAN, L., CHRISTIAN-SMITH, L. and ELLSWORTH, E. (eds) (1988) *Becoming Feminine: The Politics of Popular Culture* (London: Falmer Press).

SCHMUKLER, B. (1992) Women and the microsocial democratization of everyday life, in N. Stromquist (ed.) *Women and Education in Latin America: Knowledge, Power, and Change* (Boulder, CO: Lynne Rienner) 251–276.

SMITH, D. (1987) *The Everyday World as Problematic: A Feminist Sociology* (Boston: Northwestern University Press).

STROMQUIST, N. (1993) Sex-equity legislation in education: the state as promoter of women's rights, *Review of Educational Research*, 63(4): 379–407.

TONG, R. (1989) *Feminist Thought: A Comprehensive Introduction* (San Francisco: Westview Press).

TYACK, D. (1974) *The One Best System: A History of American Urban Education* (Cambridge, MA: Harvard University Press).

WEILER, K. (1988) *Women Teaching for Change: Gender, Class and Power* (South Hadley: MA: Bergin and Garvey).

WEXLER, P., CRICHLOW, W., KERN, J. and MARTUSEWICZ, R. (1992) *Becoming Somebody: Toward a Social Psychology of School* (London: Falmer Press).

WEXLER, S. (1982) Battered women and public policy, in E. Boneparth (ed.) *Women, Power, and Policy* (New York: Pergamon Press), 184–204.

WILLIS, P. (1977) *Learning to Labor* (New York: Columbia University Press).

PART 4
A quarter-century in perspective: implications for future research

11. Making the politics of education even more interesting

Richard G. Townsend and Norman Robinson

The time has come for studies in the politics of education to become even more interesting. The interesting sharply attacks the old truth. For instance, Lowi (1964) is interesting when he questions the notion that policy is a dependent variable of political decisions. He construes policy as an independent variable that distinctly affects the nature of subsequent politicking. Lowi's perspective-switching, now a part of political science's structure, demonstrates that the interesting can also be true and important.

With such defiance of accepted ways of thinking, the interesting elicits practical and theoretical attention (Davis 1971, Weick 1979). In celebrating heterogeneity and oppositional statements, the interesting also is postmodern in tone. It questions 'what has been taken for granted, what has been neglected' (Rosenau 1992: 8).

At bottom, we question the importance and interestingness of many truth-telling works about the political/policy context of education. They serve as little more than handy means to demonstrate our productivity. Frankly, these highly ethnocentric and catchword-prone reports are not worth the time and effort of reading onto an audio-tape for a blind colleague or of translating into another language for a friend overseas. Rather than 'proving' that charge by citing disappointing studies, however, below we select a few conclusions that climb inside surprising realities. With empirical backing, they can turn vital ideas around.

As an exemplar of what we call for, first we review iconoclastic, recent, and significant work on Max Weber ('In academe, it has long been thought . . . '). Weber has been a titan of scholarship for such a long time that we assume we know him, but that is not necessarily the case, and here we recover him as rather 'new.' Our background claim is that the interdisciplinarity, multiple levels of analysis, range of sources, methodological blending, and surprise-making of this German genius have pertinence to those concerned with politics in and around today's schools ('Any sense of Weber's limited relevance is wrong . . . '). Then, for our **foreground** theme, we find similarities between (a) the fertile contrary-mindedness of Weber and (b) nine types of propositions advanced by Davis (1970) ('Much as Weber develops interesting ideas, our inventory shows certain modern scholars . . . '). Finally, we suggest benefits would accrue if, Weber-like, more scholars were to confront established, central ways of thinking ('Further and careful negotiations of important commonplaces are recommended . . . ').

Five elements make Weber interesting and relevant today

Allusions to this 'great dead' German thinker's 7738 pages of scholarly writing are made only rarely within the politics of education subfield. In all our extensive 'modern' literature, fewer creative ideas appear than in older texts, yet we pretend that only

0268-0939/94 $10·00 © 1994 Taylor & Francis Ltd.

writings of the past 30 years are germane. In a way, this pitiful ahistoricism is curious for, well before the subfield was 'opened up' in the 1960s, Weber was politically engaged and attentive to the management of interest groups that impinge upon education. Like his successor-investigators in the politics of education, doubtless Weber's understandings of power, leadership, conflict, civic virtue, and intellectual freedom as well as his beliefs in strong constitutional democracy were influenced by his administrative experiences, his disputes with fellow-professors, and his crusades within university (as well as party and national) politics.

In academe, it has long been thought that Weber is important primarily as a marvelous structural-functional analyst of bureaucratic design. Thus he has been pictured 'as a detached, impersonal scientist, chiefly concerned with building a scientific sociology comparable to physics in its power of precise observation and abstract generalization' (Wrong 1962). One element that makes Weber germane today is the erosion of this one-sided view. Parsons (1937), his main popularizer in the West, now is being recognized for having seriously misunderstood his many-sided master. Hence, in contrast with a persona as mechanistic architect of technical rationality, Weber is garnering respect for his attention to 'cultural and historical contexts, as well as [for] the various motivations, conflicts, and influences which condition the nature of human interaction' (Samier 1993). Weber's stern critique of bureaucracy's deep structures as a dehumanizing 'iron cage' also is gaining wide attention from some who would 're-invent government' in democratic terms. (While managing a hospital during the First World War, Weber took more of a more facilitiative and human relations approach than a starchy bureaucratic one.) Thus, 75 years after his death, Weber is attracting champions, primarily European, not only for the accuracy of his insights on government, but for his profound connections of sociological, economic, legal, social psychological, religious, anthropological, philosophical, and political relationships.

A second element that now makes Weber's creative mediations interesting is that they move across all levels of analysis, from the individual, to the group, to the public agency, to the nation-state. Seldom do writers on the politics of education integrate so much, from the formal to the informal, from instrumentalities to values, from micropolitics within organizations to macropolitics of world histories. In prudently restricting their canvasses just to site or neighborhood or district or state-provincial or national affairs, they may penetrate deeply – but at the cost of closing readers off from much of the life that is in flux.

For a third element of interestingness, consider the range of Weber's sources in, for instance, his last major writing, 'Politics as a vocation' (1918/48). While current researchers commonly limit their footnotes to reports and to colleague-professors who till the same academic vineyard, Weber stunningly pulls together – within just a few short pages – sources such as Trotsky, Li Hung Chang, Emperor Max, Dostoevsky, Gladstone, Krishna, Machiavelli, the US party boss, et al. Those sources often articulate more fundamental ideas than academic specialists conventionally deliver. Weber's writing also takes on some of the poetic flavor of these artists of prose when, for instance, he warns of grim politics in a 'polar night of icy darkness and hardness.' (Times were to be darker and harder in Germany than even he could have imagined in that dawn of Hitler.)

A fourth compelling reason for our taking cues from Weber is his methodological foresight. Weber prescribed that we must be explicit about our value judgments, not hiding them behind veils of scientific competence and impartiality, and so '(v)irtually all the criticism leveled by recent philosophers of science [such as Kuhn, Feyerbend, and Popper] against logical positivism can be found in Weber's early methodological writings'

(Huff, p. 8, cited in Samier's very important study, 1993). Methodologically, Weber transcends the core duality of our era's approaches to research. His cross-cultural and cross-historical technique merits respect for its potential to bridge our era's two most distinct, strongly contesting, and uncompromising ways of knowing. Put another way, Weber's approach can help unite the 'hard' logical positivism of objective cause-and-effect explanations of behavior (which Weber calls *Erklären*) with the 'soft' interpretivism and personal meanings of hermeneutics (*Verstehen* in Weber's terms). Thus, compared with our era's protagonists of one but not both of these approaches, Weber – the mediator between objective and subjective levels of reality – commands notice not only for what he wrote but for *how* he wrote. Students of educational politics and policy today who wish to rise, interestingly, above the separateness of *Verstehen* and *Erklären* could do well to observer Weber *qua* methodologist in linking behavioral concerns to historical, institutional, and interpretive ones.

It is a fifth element of Weber, though, that deserves special honor here, his intellectual surprises. Here are two:

- In *The Protestant Ethic and The Spirit of Capitalism*, Weber repudiates the widespread understanding that a society's religion is determined by that society's economy. Taking a different tack from the Marxism of his day, Weber holds that a society's principal religion actually has an affinity for that society's economy.
- Before his major work *Economy and Society*, society's stratification system was considered monolithic. After Weber's depiction, however, stratification in certain circumstances came to be seen as comprised of three independent variables, i.e., political power, economic class, and status prestige.

From our vantage point in the 1990s, these inversions do not seem especially unexpected; we are quite accustomed to them. But at the time of inception, they were riveting.

To suggest more of what we mean by the non-interesting and the interesting, allow us to 'play with' Weber's most renowned typology. It is a trilogy that Weber returns to in his aforementioned 'Politics as a vocation.' Let us engage in a little stratification of our own:

- First, the *traditional* authority for leaders in the academic subfield of politics of education is academic success, i.e., publishing conceptual and theoretical Diamonds – as, for example, Mazzoni and Malen (1985) achieve with their article on interest-group mobilization, much as Lighthall (1989) does with his Harvest of years of work on problems, processes, and persons in school governance. In no way are these pieces Baked Alaskas, delectable and topical concoctions that are sure to perish.[1]
- Second, assuming that the politics of education subfield is a natural breeding ground for political doings and that some academic careerists will hold office therein, leaders exert *legal-rational authority* by undertaking refereeships, program chairships, presidencies, editorships, and other positions of responsibility. Wimpelberg, for one, rates statutory authority as a longtime treasurer, mobilizer, discussant, and host for the subfield's association of professors.
- Third, the resolve and capacity for inspiration at times of crisis and confusion that connote *charismatic* authority, are associated with Marshall – among other foci in presentations, she inspires followers' devotion with advocacies for more involvement by women in school governance.

This puncturing of the layman's notion that scholars are monolithic 'dry-as-dust grinds' may be gossipy, but that is not enough to qualify as intellectually interesting. The three points above are just specifications of a trilogy about authority that academics more or less assume to be helpful. Alas, no accepted view of leadership is taken apart. No breakthrough is presented, like Parson's (1937) unraveling of legal-rational authority to include expert authority.

For a new 'truth', we might assail Weber's formulations. Were we, however, to deny *all* assumptions connected to his notions of traditional, legal-rational, and charismatic authority, our argument would not be believed. At the same time, if no assumptions were denied, our rendition would be regarded as obvious, as a mere replowing of ancient grounds. Hence, to deepen understanding while maintaining credibility, an interesting assertion can only be moderate in its provocation, not too startling, not too predictable.

An impressionistic stab at that:

> Weber's distinctions help organize our understandings of authority within the politics of education academic subfield, but those distinctions miss much. Thus, at different times, Mazzoni and Malen, Lighthall, Wimpelberg, and Marshall exhibit all three kinds of thought-leadership, plus other kinds yet to be named. Instead of the three discrete kinds of authority relations that Weber isolates, there is a continuum – or perhaps there is a three-stage model where academics 'earn their spurs' with the traditional articles, then move to a legal-rational role in an academic community, and finally achieve the charismatic. Expressed differently, if an academic scribbler gives reality a new meaning, that author may become recognized which may in turn lead to being asked to officiate. Polished officiating in crisis situations may lead to new chances to produce good research, to be a gatekeeper, and warmly to personify a hitherto-neglected perspective.

Should this construction nibble at the edge of being interesting, it is because it takes on – however gingerly – the mighty Weber. In any event, any sense of Weber's limited relevance (or any seminal thinker's limited relevance) to the politics of education is wrong.

Interesting claims within the politics of education

Much as Weber develops interesting propositions, the inventory below shows certain of our modern scholars reversing everyday understandings. Similar to Davis's (1971) appraisal of general sociology, we review a number of the more familiar propositions in politics of education studies. Considerable risk attends our brief account, for necessarily we simplify findings that are complex (just as we also overlook many of the field's most valued thinkers).

Davis codifies types of contrasting propositions, and we use his same ideal types. We are not attempting a comprehensive delineation such as Weber might produce. Nor are we cataloguing paradigm shifts, much more sweeping sorts of cognitive reorientations than are interesting propositions. Aside from maintaining that the generation of interesting ideas is heuristically useful, what we do, though, is vouch that Davis's propositions can be learned and followed. That done, the politics of education subfield would not have to continue doing has been done up to now – leaving (in Davis's phrase, p. 312) 'the interesting to the inspired.'

Nine of Davis's contrasting propositions, in smaller type below, are followed by political ideas about schooling. Because of space limits, we skip Davis's three most infrequent strands of propositional pairings.[2]

The *ORGANIZATION* of phenomena:
> Type 1a: What seems to be a disorganized or unstructured phenomenon is in reality an organized phenomenon.

Davis sees the main thrust of a young ripening field as developing the Organizing Proposition of Type 1a. Before Iannaccone and Lutz's (1970) presentation of citizen Dissatisfaction Theory, the cause of the involuntary turnover of superintendents was considered manifold and fairly indeterminate. That this turnover could be related precisely to the in-migration of new residents and the election of new board members was an acute and therefore interesting attack on preceding understandings. What had seemed to be too disorganized and unstructured a process to have been logically sequenced could be codified after all.

> Type 1b: What seems to be an organized or structured phenomenon is in reality a disorganized or unstructured phenomenon.

The thrust of an old stagnating field in need of rejuvenation is to develop Disorganizing or Critical Propositions of Type 1b. The adequacies of previously accepted Organizing Propositions thereupon are called into question. Not just any Disorganizing Proposition proves interesting. What matters most are those, like Eblen's (1975–76) on the Iannaccone–Lutz Dissatisfaction framework above, that fault previous organizations of phenomena that have become widely accepted. (Eblen argues, for example, that in-migrants to a community, far from bringing new values with them, can be so attracted by the community's education that they wish for no change in programs, board members, and superintendents.)

More recently, in assessing the seemingly well-structured policy of the Reagan Administration towards excellence in education, Timar (1992) finds that Washington's funding categories strained traditional relationships, spawned competition among special interests, and eventually produced 'a fragmented and generally incoherent system of school finance' (p. 11), especially in California. Timar thus counters an almost everyday view that fiscal centralization leads to more equity and rationality in allocations.

With Shimaski in a subsequent article, Timar (1993) revisits this same leitmotif of 'what seems organized is really disorganized.' Again revealing the aggregated as disaggregated, he portrays the balance of power in California as pivoting from urban to suburban school districts. That shift sets in motion an altogether unclear political scene. In both reckonings of policy impact, Timar discerns things falling apart, leaving the murk of uncertainty.

The *COMPOSITION* of phenomena:
> Type 2a: What seem to be assorted heterogeneous phenomena are in reality composed of a single element.

The strategy here is a reductive one, looking as Plato did for the simple in the apparently complex.

Thus Peterson (1974) argues, contrary to the wisdom of that time, that school politics is not much different than other forms of local politics. In effect, the uniqueness of educational governance is an illusion and its fundamental nature lies 'below' its surface. Later, in a comparable spirit, to understand non-redistributive policies of school boards, Peterson (1981) urges us to understand the nonredistributive nature of general policies in the great cities: these cities are dependent upon and constrained by higher-level financing. Before long, Jacobs (1984) pulls together global and historical data to build a case that the function of cities, more so even than 'higher-level' nations, is economics and wealth creation. Later, Stone and Sanders (1988) adduce data to maintain that, also contrary to

Peterson, urban regimes are one of the controllers of city growth – and therefore of schools' development.

Type 2a proposition-makers show how factors not associated with each other are manifestations of a single explanatory dimension. In this mode, Weeres and Cooper (1992: 59) subsume citizens' participation in school politics as being distributed by income. Hitherto, this participation has been thought a function not only of factors such as ideological values and lengths of time in a community (Iannaccone and Lutz's factors) but also matters such as age, sex, class, public vs private regardingness, educational background, levels of urbanization, and degrees of contraction in public education (other academics' ideas). Contradicting surface impressions that each of these diverse factors has potency, Weeres and Cooper monofactorially perceive the affluent as having their educational wants satisfied privately. In their schema, the poor lack resources to participate while middle-income families evince both resources and needs to participate intensely.

> Type 2b: What seems to be a single phenomenon is in reality composed of assorted heterogeneous elements.

Here, complexity is found in the simple, and patterns often are identified only after multiple observations. Our impression is that this is the most popular variant of interestingness within the politics of education subfield.

Local politics has its share of these typologizing propositions. Among the more enduring ones, Mann (1976) concludes that school principals do not have a single orientation to their neighborhood, but have three (delegate, trustee, and politico). In the still underresearched arena we call mesopolitics, Crow (1990), Wiles et al. (1981), and Summerfield (1971) disentangle multidimensional relations between school sites and central offices. One of Summerfield's principals, for instance, deals aggressively with the community's demands – and receives 'Downtown's' resources differently from less mobilizing principals. Of late, Goldring and Rallis (1992) detail other strategies for those who would redesign their organizations and maneuver to 'fit' their environments.

Moving up the chain of command from principals, McCarty and Ramsey (1971) note that relationships between superintendents and their localities' power structures can be seen as heterogeneous – as four patterns affected by four types of community.

> The ABSTRACTION of phenomena:
> Type 3a: What seems to be an individual phenomenon is in reality a holistic phenomenon, e.g., in sociology, suicide – thought to be an individual characteristic – is in reality a societal characteristic.
> Type 3b: What seems to be a holistic phenomenon is in reality an individual phenomenon, e.g., in sociology, territoriality – thought to be a societal characteristic – is in reality an individual characteristic.

Qualifying as Type 3a propositions are assertions that delineate the 'assumptive world' and 'political culture' of state/provincial politics. What appears to be the property of an individual is portrayed as actually the property of a whole polity of which the individual is a part. In demarcating traditionalistic, moralistic, and individualist cultures and combinations thereof, Wirt (1977) lays out a far-reaching set of holistic propositions.

In particularistic propositions of Type 3b, what appears to be the property of a whole is delineated as the property of the individuals who comprise that whole. Thus, rather than discussing breakdowns in communication channels, Morris et al. (1984: 121–127) memorably define, as the 'Crazy Mother,' individuals who would improve school and society in ways that are 'off the wall' to the educational establishment. Ortiz and Ortiz (1993) have a similar individualistic focus, reviewing female Hispanic superintendents as having uniquely critical properties for moving systems toward societal equity.

Intermediate between holistic and individual foci are propositions that attend (a) to the community rather than to the society and (b) to the organization rather than to the individual. Accordingly with their look at union solidarities, Kerchner and Koppich (1993) position the locus of human phenomena at a level between the overarching polity and the particular person. Probing this intermediate context too are some of the earliest studies in the politics of local education, e.g., Hunter's Regional City (1953), Vidich and Bensman's Smalltown (1960, especially pp. 174–201), and Dahl's New Haven (1961). Perhaps someday a researcher will revisit these same sites, thereafter turning on their heads the 'old' propositions about power and socialization that these earlier scholars designed.

The *GENERALIZATION* of phenomena:
Type 4a: What seems to be a local phenomenon is in reality a general phenomenon.
Type 4b: What seems to be a general phenomenon is in reality a local phenomenon.

These too are fairly common types of propositions. Most key scholars, but not Weber, sought the generalizing proposition that encapsulated universal truth, or at least some rough approximation of it. Increasingly now, with the outlook that elements can vary considerably among different populations, more openness exists for propositions that are localizing.

Thus with rigorous procedures, based on a two and a half-year study of 12 high schools in 11 states, Weiss (1993) puts into a general and national context what professionals likely know about a nearby school or district's shared decision making (SDM). SDM may give teachers some influence and ownership over limited issues, but so far it has not especially led teachers to 'stress innovation nor craft creative strategies to improve student achievement' (p. 89).

Teacher commitment may be regarded as local and idiosyncratic until a persuasive countrywide literature review on that subject, (e.g., Firestone and Pennell 1993) comes along to weave together findings, logical deductions, evaluations, and caveats to confirm, challenge, and extend readers' outlooks. Firestone and Pennell hence offer an interesting proposition when they argue that formative feedback – which 'everyone knows' as characterizing the supervision process – also characterizes teacher commitment. Prior to that review, effective feedback had not been much linked to commitment, and so Firestone and Pennell somewhat 'catch' us by asking that we 'imagine how schools can be reconfigured to provide teachers' (p. 519) with supervisory feedback.

All of us have understandings that we take for granted, and so a denial of the localizing proposition, along with an emphasis on variation, can prove interesting too. Thus, through her negation of a one-size-fits-all approach to home–school relations, Graue (1993) shows that certain behavioral or policy characterizations hold for some parental groups but not for others. Unlike Europeans, American scholars in the politics of education today largely ignore the Russian Vygotsky, but not Graue. Through that Russian's framework, she examines the language and actions of parents to illustrate how local contexts can set up assorted kinds of home–school relations.

The *STABILIZATION* of phenomena:
Type 5a: What seems to be a stable and unchanging phenomenon is in reality an unstable and changing phenomenon.
Type 5b: What seems to be an unstable and changing phenomenon is in reality a stable and unchanging phenomenon.

Fortunately the subfield has beckoned researchers with the historian's penchant for sifting through data on temporal continuities or lacks thereof. Rogers and Chung (1983) on New York, Schwartz on Alberta (1986), Rury, Cassell *et al.* on Milwaukee (1992), McNeese on Salt Lake City (1992), and Mirel on Detroit (1993) are among those who work effectively tracing change and stability.

For a Type 5b proposition, Wong (1992) finds few 'choice programs in the midst of powerful central bureaucracy in the urban district.' Wong meanwhile observes 'dedicated parents vigorously involved in [new] site-level governing councils while others continue to exit to the nearby district or the non-public sector.' Overall, new governing structures co-exist with old mechanisms, yet, in Wong's view, the remaining stabilities have strong advantages. That is, the still-powerful central bureaucracy of a city possesses institutional memory, policy enforcement capacity, ability to conduct program evaluation, and coordinative authority for mediating rival interests.

In the exposé mode (not done enough in our field), Madsen (1994) also points to unyieldingness in the face of mandated reform. That schools eat change agents for breakfast does not surprise the professional. Still, in this era of 'break-the-mold' schools in the USA, a state department of education is 'supposed' to have a different aura, at least carrying forward a fresh perspective. Lest laypersons be so rash as to believe all state legislators and functionaries, however, Madsen's behind-the-scenes vignettes specify how 'new' programs are neutered these days. (What is interesting to a professional and to a layperson can differ.)

The *FUNCTION* of phenomena:
> Type 6a: What seems to be a phenomenon that functions ineffectively as a means for the attainment of an
>   end is in reality a phenomenon that functions effectively.

One example: through a study of desegregation in 119 districts, Rossell (1990) refutes the convention that mandatory reassignment is the only way to achieve school desegregation. More effectively than usually assumed, voluntary plans function in facilitating students' racial integration, she concludes. Indeed, compared with mandatory plans, voluntary plans are judged superior in equity, efficiency, and effectiveness – which also is not what seasoned civil rights proponents have anticipated, given their predilection to liberals' strategy of court rulings to advance their interests.

Such a Type 6a proposition can involve rejecting a *manifest* (i.e., a generally apparent) dysfunction within a phenomenon and then finding a *latent* (i.e., not generally apparent) redeeming function therein. In an analysis of causes for a wholesale reform of school governance in Chicago, Hess (1993) argues that it was the liberal *strategies* (a manifest function) that failed. Defeat did not come, Hess says in contra-distinction to common belief, because of the liberal *perspective* of bringing extensive political changes peacefully (a latent function).

> Type 6b: What seems to be a phenomenon that functions effectively as a means for the attainment of an end
>   is in reality a phenomenon that functions ineffectively.

Sykes (1987) points to the pathological adaptations teachers make to personnel incentives – certain 'rewards' turn out to be disincentives.[3] But, more typically in the literature, local boards of education seem to exemplify this surprise. Elected boards may have developed originally to provide for lay control, but Zeigler and Jennings (1974) uncover scant evidence that these bodies actually usher in 'responsive and responsible governance.' Rather, electoral 'competition is limited, with sponsorship and pre-emptive appointments common. Opponents of the status quo are infrequent; incumbents are rarely challenged and more rarely still defeated' (p. 245). Whenever superintendents are ensconced as dominant figures, board members become spokespersons to the community for those superintendents, no longer brokering grassroots concerns. Granted, a fair amount of ink has been spilt in academic journals contesting this proposition, yet the view of democratic lay control of education by boards has been questioned, interestingly and importantly.

Type 6b kinds of propositions can be radical. An implication is that we ought to try changing or even abandoning political institutions that have destructive consequences. Chubb and Moe (1990) lately refute the Zeigler and Jennings proposition, insisting that a large part of what prevents public schools from fostering high achievement has been this very element of democratic control. From research on two districts, Hannaway (1993) yields a more subtle finding: contrary to the standard thinking that organizational decentralization expands the discretion of those closest to the problem, teachers in decentralized schools appear to have less discretion than those in traditional settings.

> The EVALUATION of phenomena:
> Type 7a: What seems to be a bad phenomenon is in reality a good phenomenon.
> Type 7b: What seems to be a good phenomenon is in reality a bad phenomenon.

To bolster the dismal rating of an activity, the researcher can choose indicators that 'authenticate' the brightness of the phenomenon as a whole. Accordingly, if her or his objective is to offset a view that schools are awful, the maker of interesting propositions can scout out indicators that emphasize achievements. Second, the researcher with reason to be gloomy about today's schools can try to shift the rating of a phenomenon by shifting the standard with which it is being compared. This sort of skew of consciousness is essentially what Chubb and Moe (1990) do in moving the public schools' comparison base to better-achieving, market-oriented, non-public schools that do not have to respond to democratic pressures.

Of course, for their part, analysts inclined to undercut their audiences' appraisals of schools as lackluster can muster comparisons too. Comparisons can be to (a) other societies' schools, (b) schools' past history, (c) schools' potential futures, or (d) some edition of Utopia. Comparisons of course can be negative as well as positive. Thus Scheurich (1993) compares coordinated children's services with the violence inflicted on poor minorities. Scheurich's point in short is that coordination by professionals of other people's lives, rather than enabling, is patronizing and disempowering.

If various partnerships presently are taken at face value, business input is a 'good' for education, leveraging money and expertise to meet identified school and community needs (McGuire, 1990). With an eye on lower-income groups, however, Spring (1992) as well as Borman *et al.* (1993) are among those who demur. Such critics argue that 'overwhelmingly, business interest and involvement in education reform have been compelled by narrow business self-interest, contradictory to the interests of women, people of color and children . . . . Little optimism for future business-led efforts to generate structural change in education can be expected' (Borman *et al.* 1993: 69). In making that case, these anti-business critics chose negative indicators, shifting the standard away from workplace productivity. Potential inequities for students are cast as flowing out of the new partnerships.

> The CO-RELATION of phenomena:
> Type 8a: What seem to be related or interdependent phenomena are in reality uncorrelated or independent phenomena.
> Type 8b: What seem to be unrelated or independent phenomena are in reality correlated or interdependent phenomena.

The impact of schools on children has been the subject of both sorts of co-relational propositions. Until 1979 or so, an accepted view among a number of high-profile scholars was that public schools, in the end, appear not to have any impact upon class and ethnic differences in students' achievements (Coleman *et al.* 1966, Jencks 1972, Plowden Report 1967). In due course, however, Rutter *et al.* (1979) proposed otherwise, showing that it did matter which secondary school a child attended. Giving impetus to the effective school movement, these Britons also established that the way a school is organized is crucial to setting conditions for learning.

The *CO-EXISTENCE* of phenomena:
Type 9a: What seem to be a phenomena which can exist together are in reality phenomena which cannot exist together.
Type 9b: What seem to be a phenomena which cannot exist together are in reality phenomena which can exist together,

These types of interesting propositions are relatively rare, since few phenomena are so incompatible with one another that they completely negate the other's existence.

British research, though, again provides a quick example. England's Educational Reform Act of 1988 abolished restrictions on attendance boundaries, thereby making it possible for parents to choose freely which schools their children would attend. Ostensibly, schools would be obliged to accept any child provided the school had space. The principle of free choice was designed to give all parents an equal opportunity. Legislators saw choice and equity initially as compatible co-equal objectives of the Act. Over time, however, choice and equity do not seem to be able to co-exist comfortably. Hence, in those well-endowed schools which typically have more applicants than spaces, school authorities regularly exclude children with a history of emotional or behavioral problems (Woods 1992). These 'problem' children are then forced to seek places in less favored schools. To a degree, then, choice tends to produce inequity.

For a North American example, Rallis (1993) proposes that elected school boards and school restructuring may be a contradiction in terms. She observes that the innovation necessary to move American education 'back on track' is antithetical to the operational styles of most board members.

Examples of Type 9b propositions occur in the oxymorons of commentators on politics and policy in education. These include dynamic ambiguities such as 'organized anarchy,' 'loose coupling,' and 'principled compromise.' Qualities that seem incongruous or contradictory can co-exist, maybe even with satisfactory outcomes.

The *IMPORTANCE* of phenomena:

To suggest that Davis (1971) has not fully mapped the domain of what is interesting (as well as important and true), we invoke a pair of propositions that he missed. First:

a: What seem to be important concerns are in reality not centrally important.

Thus, with great faith in Second Chance programs, Levin (1992) musters an intriguing idea that undercuts received wisdom: for good reasons, he does not see the dropout issue as one which schools need to be importantly concerned about. Second:

b: What seem to be unimportant concerns are in reality phenomena which are centrally important.

Up until now, the literary style in which a policy is written has hardly been considered critical. Woods (1994), however, recently has claimed that the wording and presentation of a policy affects teachers' engagement and subsequent support of the policy involved. Thus at a time when various mainstream scholars insist that institutions matter, leadership matters, external resources matter, and implementation techniques matter, one small voice draws upon reader-response theory to suggest that (of all things) liveliness in writing matters.

Any number of other types of interesting propositions is possible. For instance, a scholar could write about well-accepted concepts that are unworkably narrow or confusingly wide, or reframe discussions from the overly technical to the moral.

Beyond interestingness to paradoxes, Weber, and more

In a highly selective account, we have pointed to a handful of seemingly contradictory assertions in studies of the politics and policy of education. While not exhaustive, our search has been sufficient to suggest that the subfield is not riddled by groupthink. Yet because we have had to 'dig' to find these seeming contradictions, our hunch is that many articles in this subfield have not been as interesting as they might have been. Either that, or authors in our field commonly do not call readers' attention to the new departures they take.

Authors of some of those and more substantive articles may rebut, either arguing that (a) the world is more gray and less black-and-white than contrasting propositions presume, (b) the contradictions we seek would only ruffle people of action who crave the decisiveness of certainty, and (c) a lack of consensus within the scholarly community leads to the public's low estimation of educational research. Also, we may be told that the interest-monger hedonistically craves novelty and wild surmise while lamely forgetting that many find security in denying the interesting. The rejection of garden-variety understandings is only an intellectual game, critics also may say (perhaps by way of advancing research games of their own).

Notwithstanding these dissents to our theme, we see a cultivation of the interesting as a call for flexibility and for interim understandings. 'Weigh your own and others' assumptions, even at the risk temporarily of putting the research community into disarray,' we see Weber, Davis, and ourselves as saying, 'and always examine conflicting or opposite hypotheses. Move beyond (1) the verifying of accepted ideas and (2) the unique vision that lacks rigor. With your propositions, have an openness to assorted realities and to notions that differ from your own. So, if you want to help establish truth, keep thinking.' That strikes us as fair advice for a subfield 25 years mature. Further and careful negations of important commonplaces are commended.

Although Davis does not use the word 'paradox' to describe the import of his search of the contrasting view, his whole approach stimulates us to do what paradoxes do – plant doubts and keep us open to possibilities, including the central political thought that understanding does not necessarily come in the form of unequivocal answers (Tinder 1991: 21). Not for nothing is the patron of politics, Janus, two-faced. In our experiences, truth in the intricacies of politics and policy, and in education too, rarely is confined to one universally compelling proposition. As Milton notes in *Areopagitica*, where there is knowledge in the making there of necessity will be much arguing, much writing, many opinions.

Indeed, the 'either–or' alternative that marks western thinking might be transcended by a 'both–and' resolution that reveals the interdependence of qualities upon their opposites (Barlosky 1994). As students of Tao have remarked, without 'off' there is no 'on.' Without 'dark,' there is no 'light.' Without 'messiness,' there is no 'order.' Difficult and easy complete one another. High and low determine one another. Front and back give sequence to one another (Lao Tze 1958: 143).[4]

What will this sort of ying-yang thinking do to the subfield? One answer can be found in that final 1918 lecture of Weber's in Munich. Weber evokes two propositions as being played out by those in the vocation of politics:

1. He characterizes *an ethic of ultimate ends* with fiercely held personal principles. Elsewhere, Weber alludes to this commitment to a cause as a *passion* to reform a corrupt world, no matter what the consequences.

2. As a 'guiding star,' Weber interestingly and contrarily-mindedly invokes an *ethic of personal responsibility* to the cause of politics. In that ethic, purities of principle are tempered pragmatically. There, politicians soberly act with regard to the consequences of their actions and of the world's actions.

In Weber's construction, both contrasting ethics – the ethic of passion as well as the ethic of personal responsibility – are necessary.[5] But how to overcome the seeming duality, and other dualities as well?

Weber summons a mediating force between individuals' autonomy and responsibility. He defines this force, this 'decisive psychological quality' allowing calmness and concentration, as *proportion and perspective*. Weber holds that somehow the politician (and the student of educational politics, we would add):

> ... reaches the point where he says: 'Here I stand: I can do no other.' Insofar as this is true, an ethic of ultimate ends and ethic of responsibility are not absolute contrasts but rather supplements which only in unison constitute a genuine man – a man who *can* have the 'calling for politics'. (1918/48: 127)

Thus, after agonizing over two paradoxically opposing ethics, Weber points to room for both. Anticipating the next decade's work by Mary Parker Follett on integrative solutions, Weber urges politicians to take a stand by fusing contradictions.

Unlike some behavioral scientists, Weber does not quantify the desired portions between passion and responsibility, nor does he lay out analytic decision trees to help us reach accommodations. Yet he leaves no doubt that he expects politicians to piece together proportional adjustments where the ethics of ultimate ends and of responsible consequences harmonize with each other.

In an overview like this for one subfield, certainly it is never too late to allude to the perennial debate about the purpose of the subfield's scholarship. In that debate, we see oppositional ethics similar to those that Weber marked. That is, we see a passion for theory and for the subfield taking on more of the dimensions of a cognate and theory-building discipline; Burlingame (1987) is only one of the more explicit spokespersons for this pro-understanding view. We also observe a strong ethic involving personal responsibility. This disposition toward statecraft responds to cries of administrators, reformers, and others who demand practical help today in solving political problems engulfing schools (Sroufe 1980). We would be exaggerating to suggest that the subfield might experience a polar night of icy darkness over this and other tensions (e.g., *Erklären* vs *Verstehen*), but esteemed writers and speakers on panels often seem to talk past each other, as if little chance for compromise existed between aspects of the field.

Through the crossfire that such contesting positions promote, those with a calling to study and learn from the politics of education might be stimulated to move on to more paradoxical thinking. Those with this calling might reach for, and achieve, that very proportion and unison that Weber values. Thereafter, who knows? Their balances might turn out to have more wisdom than what is strictly interesting.

Notes

1. Diamonds, Harvests, and Baked Alaskas are terms that Vandermeulen (1972) uses to classify manuscripts submitted to an economics journal.
2. Hence, we do not deal with: the *CO-VARIATION* of phenomena (i.e., what seems to be a positive co-variation between phenomena is in reality a negative co-variation, and what seems to be a negative co-variation is in reality a positive one); the *OPPOSITION* of phenomena (i.e., what seem to be similar or nearly identical phenomena are in reality opposite, and vice versa); and the *CAUSATION* of

phenomena (i.e., what seems to be the independent variable in a causal relation is in reality the dependent phenomenon, and vice versa.)

3. For this reference, we are obliged to Carol Bartell.

4. An example of this transcending quality in the politics of education subfield is the insight, attributed to Kaufman, that centralization creates countervailing needs for decentralization, and vice versa. So it is too that administrators appreciate that they often have to change just to preserve their schools' stabilities. Paradoxically too, authorities who make political appointments appreciate that compelling arguments can be made for choosing representative *or* unrepresentative groups as well as for picking groups that represent the whole society *or* just the clientele. Finally, students of policy realize that behind every policy goal (e.g., equality) are conflicting but plausible conceptions of that goal (e.g., equality in process vs equality in distribution).

5. Weber's openness is similar to that of physicists who made scientific advances at Cambridge in the 1930s: on Mondays, Wednesdays, and Fridays, they applied wave theory to explain light; then on Tuesdays, Thursdays, and Saturdays, they adapted the seemingly irreconcilable newer photon/particle model of quantum dynamics. Paradoxically, physicists ultimately determined that both forces, wave as well as particle, are at play.

References

BARLOSKY, M. (1992) East meeting West: the case of Alan W. Watts – alternative foundations for inquiry and practice in educational administration, paper presented at conference on Universities East and West, OISE.

BORMAN, K., CASTENELL, L. and GALLAGHER, K. (1993) Business involvement in school reform: The rise of the business roundtable, in C. Marshall (ed.) *The New Politics of Race and Gender* (London: Falmer Press), 69–83.

BURLINGAME, M. (1987) The shambles of local politics, *Politics of Education Buletin*, 13(3), 3–8.

CHUBB, J. and MOE, T. (1990) *Politics, Markets, and America's Schools* (Washington, DC: Brookings Institutions).

COLEMAN, J. S. *et al.* (1966) *Equality of Educational Opportunity* (Washington, DC: US Department of Health, Education, and Welfare).

CROW, G. (1990) Central office influence on the principal's relationship with teachers, *Administrator's Notebook* 34(1), 1–4.

DAHL, R. A. (1961) *Who Governs?* (New Haven: Yale University Press).

DAVIS, M. (1971) That's interesting! Towards a phenomenology of sociology and a sociology of phenomenology, *Philosophy of Social Science*, 1, 309–344.

EBLEN, D. R. (1975/76) Local-school district politics: a re-assessment of the Iannaccone and Lutz model, *Administrator's Notebook*, 24, 1–4.

FIRESTONE, W. A. and PENNELL, J. R. (1993) Teacher commitment, working conditions, and differential incentive policies, *Review of Educational Research*, 63(4): 489–525.

GOLDRING, E. B. and RALLIS, S. (1992) Principals as environmental leaders, paper presented at the annual meeting of the University Council of Educational Administration.

GRAUE, E. (19983) Social networks and home-school relations, *Educational Policy*, 7(4), 483–490.

HANNAWAY, J. (1993) Decentralization in two school districts: challenging the standard paradigm, in J. Hannaway and M. Carnoy (eds) *Decentralization and School Improvement: Can We Fulfill the promise?* (San Francisco: Jossey-Bass), 135–162.

HESS, G. A. Jr (1993) Race and the liberal perspective in Chicago school reform, in C. Marshall (ed.) *The New Politics of Race and Gender* (London: Falmer Press), 85–96.

HUFF, T. (1984) *Max Weber and the Methodology of the Social Sciences* (New Brunswick, NJ: Transaction Books).

HUNTER, F. (1953) *Community Power Structure* (Chapel Hill: University of North Carolina Press).

IANNACCONE, L. and LUTZ, F. W. (1970) *Politics, Power, and Policy* (Columbus: Charles E. Merrill).

JACOBS, J. (1984) *Cities and the Wealth of Nations: Principles of Economic Life* (New Tork: Random House).

JENCKS, C. (1972) *Inequality* (New York: Basic Books).

KERCHNER, C. and KOPPICH, J. (1993) *A Union of Professionals* (New York: Teachers College Press).

LAO TZE (1958) The way and its power: a study of the Tao To Ching and its place in Chinese thought, trans. A. Waley (New York: Grove Press.

LEVIN, B. (1992) Dealing with dropouts in Canadian education, *Curriculum Inquiry*, 22(3): 257–270.

LIGHTHALL, F. with ALLEN, S. (1989) *Local Realities, Local Adaptations: Problems, Process, and Person in a School's Governance* (London: Falmer Press).

LOWI, T. (1964) American business, public policy, case studies, and political theory, *World Politics*, 16(4): 677–715.

MADSEN, J. (1994) *Educational Reform at the State Level: The Politics and Problems of Implementation* (Washington: Falmer Press).

MANN, D. (1976) *Politics of Administrative Representation* (Lexington: Lexington Books).

MAZZONI, T. and MALEN, B. (1985) Mobilizing constituency pressure to influence state education policymaking, *Educational Administration Quarterly*, 21(1): 91–116.

McCARTY, D. J. and RAMSEY, C. E. (1971) *The School Managers: Power and Conflict in American Education* (Westport: Greenwood).

McGUIRE, K. (1990) Business involvement in the 1990's, in D. E. Mitchell and Goertz, M. E. (eds) *Education Politics for the New Century* (London: Falmer Press), 107–117.

McNEESE, P. A. (1992) The process of decentralizing conflict and maintaining stability: site council enactment, implementation, operations, and impacts on the Salt Lake City School District, 1970–1985, unpublished PhD dissertation, University of Utah.

MIREL, J. (1993) *The Rise and Fall of an Urban School System: Detroit, 1907–81* (Ann Arbor: University of Michigan Press).

MORRIS, V. C., CROWSON, R. L., PORTER-GEHRIE, C. and HURWITZ, Jr E. (1984) *Principals in Action* (Columbus: Charles Merrill).

ORTIZ, F. I. and ORTIZ, D. J. (1993) Publicizing executive action: the case of Hispanic female superintendents, in C. Marshall (ed.) *The New Politics of Race and Gender* (London: Falmer Press), 155–167.

PARSONS, T. (1937) *The Structure of Social Action* (New York: Free Press).

PETERSON, P. (1974) The politics of American education, in F. Kerlinger and J. Carroll (eds) *Review of Research in Education, 2* (Itasca: Peacock), 348–389.

PETERSON, P. (1981) *City Limits* (Chicago: University of Chicago Press).

PLOWDEN REPORT (1967) *Children and their Primary Schools*. (London: HMSO).

RALLIS, S. with CRISCO, J. (1993) School boards and school restructuring: a contradiction in terms?, paper presented at annual convention of American Educational Research Association.

ROGERS, D. and CHUNG, N. (1983) *110 Livingston Street Revisited: Decentralization in Action* (New York: New York University Press).

ROSENAU, P. M. (1992) *Post-modernism and the Social Sciences: Insights, Inroads, and Intrusions* (Princeton: Princeton University Press).

ROSSELL, C. (1990) *The Carrot or the Stick for School Desegregation Policy* (Philadelphia: Temple University Press).

RURY, J. L. and CASSELL, F. (eds) (1992) *Seeds of Crisis: Public Education in Milwaukee since 1920* (Madison: University of Wisconsin Press).

RUTTER, M., MAUGHAN, B., MORTIMORE, P. and OUSTON, J. (1979) *Fifteen Thousand Hours: Secondary Schools and their Effects on Children* (Cambridge: Harvard University Press).

SAMIER, E. A. (1993) A study of the relevance of Max Weber's work to educational administration theory, unpublished PhD. dissertation, University of Victoria, Canada.

SCHEURICH, J. J. (1993) Policy archaeology: a postmodernist approach to policy studies, paper presented at annual conference of the University Council for Educational Administration.

SCHWARTZ, A. (1986) Teaching hatred: the politics and morality of Canada's Keegstra affair, *Canadian and International Education*, 16(2): 5–28.

SPRING, J. (1992) Knowledge and power in research into the politics of urban education, in J. G. Cibulka, R. J. Reed and K. Wong (eds) *The Politics of Urban Education in the United States* (London: Falmer Press), 45–55.

SROUFE, G. (1980) The very last word about politics of education research, *Politics of Education Bulletin*, 9(3): 3–6.

STONE, C. and SANDERS, H. (eds) (1988) *The Politics of Urban Development* (Lawrence: University Press of Kansas).

SUMMERFIELD, H. L. (1971) *Neighborhood-based Politics of Education* (Columbus: Charles E. Merrill).

SYKES, G. (1987) Teaching incentives: constraint and variety, *Proceedings of a Seminar on Incentives that Enhance the Teaching Profession* (Elmhurst: NCREL), 57–86.

TIMAR, T. (1992) Urban politics and state school finance in the 1980s, in J. G. Cibulka *et al.* (eds) *The Politics of Urban Education in the United States* (London: Falmer Press), pp. 105–22.

TIMAR, T. and SHIMASAKI, D. (1993) Categorical wars: zero-sum politics and school finance, in C. Marshall (ed.) *The New Politics of Race and Gender* (London: Falmer Press), 19–35.

TINDER, G. (1991) *Political Thinking: The Perennial Questions* (Boston: HarperCollins).

VANDERMEULEN, A. (1972) Manuscripts in the maelstrom: a theory of the editorial process, *Public Choice*, xii, 107–111.

VIDICH, A. J. and BENSMAN, J. (1960) *Small Town in Mass Society: Class, Power, and Religion in a Rural Community* (Garden City: Anchor Books).

WEBER, M. (1918/48) Politics as a vocation, in H. Gerth and C. W. Mills (eds) *From Max Weber* (London: Routledge and Kegan Paul).

WEERES, J. G. and COOPER, B. (1992) Public choice perspectives on urban schools, in J. G. Cibulka, R. J. Reed and K. Wong (eds) *The Politics of Urban Education in the United States* (London: Falmer Press), 57–69.

WEICK, K. (1979) *The Social Psychology of Organizing* (Reading: Addison).

WEISS, C. (1993) Shared decision making about what? A comparison of schools with and without teacher participation, *Teachers College Record*, 95(1), 69–92.

WILES, D., WILES, J. and BONDI, J. (1981) *Practical Politics for School Administrators* (Boston: Allyn and Bacon).

WIRT, F. M. (1977) State policy culture and state decentralization, in J. Scribner (ed.) *Politics of Education* (Chicago: National Society for the Study of Education), 165–187.

WOODS, L. (1994) The implications of engagement in teachers' responses to educational policy documents, unpublished EdD dissertation, OISE.

WOODS, P. (1992) Responding to the consumer: parental choice and school effectiveness, paper presented at the International Congress for School Effectiveness and Improvement (Victoria, BC).

WONG, K. K. (1992) The politics of urban education as a field of study: an interpretive analysis, in J. G. Cibulka, R. J. Reed and K. Wong (eds) *The Politics of Urban Education in the United States* (London: Falmer Press), 57–69.

WRONG, D. (1962) Max Weber, in *Architects of Modern Thought* (Toronto: Canadian Broadcasting Corporation), 145–226.

ZEIGLER, L. H. and JENNINGS, M. K. (1974) *Governing American Schools* (North Scituate: Duxbury).

12. *Educational politics and policy: and the game goes on*

Jay D. Scribner, Pedro Reyes and Lance D. Fusarelli

> The game analogy not only makes the political process easier to understand, it also draws attention to the dynamic nature of the process. A game, after all, is not static but full of action. So it is with the game of American politics, for the action is nonstop. (Frantzich and Percy 1994: 3)

This commemorative yearbook explores the competing values that shape the rules of the game, examines the high-stakes political arenas in which the educational politics game is played, and illuminates the policy outcomes and consequences critical to the education of children. *The Study of Educational Politics* challenges us to venture forth into uncharted territories, beyond the classical paradigm that some say inspired us to adopt a bureaucratic view of a game where the players behave as cogs and wheels within precisely defined game-board boundaries, while conforming to strictly enforced rules. Games played on the constantly changing playing fields of the politics of education are rarely this boring. They are highly contested, continuously in motion, and played on fields where boundaries are obscured and rules blurred by the exigencies of social change. In using the game analogy to conclude the yearbook, we draw heavily on our colleagues' analyses of value conflicts within political arenas where power is the medium through which players, rules, strategies, and outcomes affect all those accountable for our education in our society.

We examine content themes addressed throughout the yearbook while simultaneously considering trends over the past quarter century that may foreshadow new directions for research in the 21st century. For instance, we ask questions about the impact playing-field conditions have on whether an educational reform persists or not. We offer examples of ways educational politics and policy games are played. Finally, we recognize the disillusionment prevalent in educational research with mechanistic models of human behavior and the attendant problem of values, universals and objectivity. We advocate diverse ways to study educational politics and policy games. Our position supports theoretical and methodological pluralism in the politics of education, a stance that embraces a holistic approach inclusive of various theoretical perspectives and methodological approaches which have been largely overlooked in inquiries into the politics of education.

Changing playing fields and educational reform

The nature of the playing field is determined by where the game of educational politics is played, how it is played, and who wins and loses. Playing fields are characterized simultaneously by relatively stable governmental jurisdictions, by established communities and other similar territorial dimensions, and by constantly shifting cultural traditions, norms, expectations and values. Reforms come and go in the United States despite institutional continuity because of changing playing conditions resulting from recurring

0268–0939/94 $10·00 © 1994 Taylor & Francis Ltd.

conflicts over fundamental values in schools and society at large. Knowledge of the conditions under which the game of educational politics is played assists us in understanding the focus of education reform.

At the outset of this yearbook, Stout, Tallerico and Scribner assume that playing fields generate conflict and that those conflicts are unresolvable in a pluralist democratic system because they reflect tensions among competing values. This contention is well documented in the literature (Cuban 1990, Guthrie 1987, Iannaccone 1967, Marshall *et al.* 1989). Within our political culture, the inevitable contests over contemporary issues taking place in political arenas at all levels of school governance absorb our immediate attention. Yet, over the long term, as a relatively common set of rules for playing the politics of education game passes from one generation to the next, ideologies persist; conservative and liberal values appear, reappear and appear again (Burnham 1970, Cuban 1990, Schlesinger 1986).

Thus, playing fields change as individuals move in and out of political arenas. When conservative values dominate the playing fields, the players concern themselves with high academic standards, orderliness, efficiency, and productivity; when more liberal values dominate, issues concerning equity, student access to programs, linking schools to work in the community, and reducing academic achievement gaps between student groups are emphasized.

One of the principal conflicts generated by this tension over competing values concerns the role of ideology in education. Ideological belief patterns are expressed in terms of how individual players apply their personal experiences to the decisions and policies made in political arenas (i.e., in schools, districts, state and federal bureaucracies), and, likewise, how players with similar beliefs and attitudes view such issues as equality, the purpose and size of government, the inevitability of change or the virtues of stability and order. These periodic shifts, as Mazzoni suggests, have been 'energized' by dramatic events, new leadership, media attention and the bully pulpit. The playing fields differ among the different states as evidenced by contrasts in values and attitudes associated with the highly contested battles over school finance reform, school choice, school desegregation, and religion in the schools.

Playing fields change because of individuals who are in a position to wield influence over other players. As Sroufe asserts, the players themselves bring personal experiences and values to the playing fields that shape their agendas and the nature of the game they wish to play. Among his examples are President Lyndon B. Johnson, a former educator who created our nation's largest educational program, and more recently Senator Tom Harkin whose influence on the passage of bills concerning technical assistance for the handicapped undoubtedly grew out of his first-hand experience with a brother with a hearing disability. Throughout this volume several writers have argued that the players, including school administrators, legislators, governmental leaders, community élites, interest group leaders, and the like bring their own predispositions, *visions*, and values to the political arena as they seek to control agendas and determine outcomes, manage bureaucratic myths and their own images, gain control of real and symbolic resources, and manipulate implementation processes.

Playing fields also change largely because of the prevailing ideological preferences of the times as has been expressed, for example, through educational reforms enacted during the administrations of the seven United States Presidencies since the inception of the Politics of Education Association. Administrations in Washington, DC representing different ideologies can be distinguished between those who have emphasized equality of educational opportunity and fiscal and academic equity (social justice), and those who have seen government in a more limited role, placing high priority on orderliness, efficiency, and productivity.

Cibulka makes it abundantly clear that these distinctions have in recent years engendered not only 'politicized policy-oriented research, but academic think tanks (in Washington, DC and elsewhere) representing conservative or liberal values' that use their 'political resources' to influence what is studied, and how public opinion is to be manipulated by 'deliberately slanted research.' Interestingly, Wong suggests that the recent resurgence of a federal interest in education backed by some modest improvements in amounts and redistributions of resources under the Clinton administration breaks the mold by combining two competing values, 'achieving equity and efficiency simultaneously.'

In sum, during this first quarter century of the evolution of the politics of education as a field of study, educational policy has moved from an earlier call for school district consolidation with fewer people making political decisions and allocating values for more people to the recent trend toward more inclusive, participatory (site-based) decision making.

This shift in emphasis, however, is far from permanent. Iannaccone and Lutz remind us that because of shifting power bases and changing playing fields, local control has become a myth and representativeness of diverse populations a fiction. Malen further submits that within schools where instructional and curriculum decisions are made tensions exist over who has the legitimate role and right to make decisions in particular areas. There is a great dispute among educators as to whether the recent emphasis on restructuring through shared decision making will produce a lasting change in educational governance or whether it represents merely a pause in the movement toward ever greater state and federal involvement in education.

Educational reform emanates from persisting conflicts over unresolvable values. Understanding how playing fields and political arenas influence the game is crucial to understanding educational reform. Playing fields are in perpetual motion, and political arenas are constantly shifting boundaries where governing powers between national, state and local units of government are shared, and the outcomes of the games played are shaped by values emanating from cultural, historical, and technological changes. As indicated earlier, Wong claims that since the inception of the politics of education field reform efforts have tended to oscillate between two perspectives that can be roughly placed on 'an "equity–efficiency" continuum.' These two perspectives create tensions and, thus, winners and losers.

Educational policy games: winners and losers

Because tensions between competing social values are ever present in schools, they rise to the surface when external events trigger individuals and groups to voice policy differences and demand change in schools. These dilemmas require political negotiation and compromises among policy makers and interest groups. There is no solution to these conflicts; there are only political trade-offs. In this sense, as Sroufe suggests, the game never ends. These conflicts and political trade-offs are what make it difficult to determine the winners and losers in educational politics games. Sometimes only one player or group of players win, in other situations more than one player wins, and in still others most of the players lose.

That winners and losers are difficult to determine is a theme found in many of the yearbook chapters. For example, those concerned with mapping the strategies and tactics used in determining or predicting winners and losers have adopted a variety of analytical models. Boyd, Crowson and van Geel offer game theory as an example of a rational choice theory that can be used to predict winners and losers of political games. Zero-sum games, positive-sum games, mixed-motive games, prisoner's dilemma, chicken, and convergence games are offered as ways to play the politics of education game. Policy researchers systematically sort out the logical effects changes in one strategy (variable) will have on other factors in an imminent contest.

Cibulka offers disquieting evidence about the outcome of macropolitical games being decided by the accuracy and thoroughness of information available to the players. He argues that one can indeed determine winners and losers. Policy élites most typically win because they control information. As players, they are fewer in number and active in as many arenas as possible. Their primary strategy is to control agendas so as to manage the flow of information and minimize outside influence. Outcomes of micropolitical games, however, appear to be in the hands of the principal who controls information, sanctions, and rewards. Malen asserts that both teachers and principals use strategies of every description to protect their turf, and 'avert the sanctions and contentious exchanges' that threaten traditional power relationships. Depending on how the game is played, teachers will either withdraw from the playing field into the classroom arena, or collaborate, compromise, and form positively directed coalitions in the workplace.

The movement toward the privatization of public schooling offers another kind of educational policy game. Leading examples of these policy games are the charter schools and school choice movements, movements which are gaining increasing momentum in several states, as well as the national level. In these games the traditional winners are the private organizations that take over the responsibility for schooling. Potential losers in this game are school professionals and possibly a large segment of low-income families and their children (Boyd and Kerchner 1988).

The distinction between 'public' and 'private' has traditionally occupied an important place in our political and legal notions. During the 'Reagan revolution', however, the public and private distinction became less well defined, perhaps even inverted (Wolin 1989). A concerted attempt was made to make a profitable business out of social problems previously thought to be in the public domain (Brooks et al. 1984). Given recent political trends, this blurring of distinction between public and private spheres will likely continue well into the next century.

The appropriation of public goals by private enterprise suggests that state power via the policy arena is being expanded, but no longer flowing from a common center (Wolin 1989). If this contention is correct, then the players of this educational policy game will have to consider the possibility that policies such as site-based management and decentralization, which are ostensibly offered as tools to provide more local control and freedom of choice to individual schools, will produce opposite effects. These policies may only serve to strengthen the involvement of the 'state' in local affairs. Further, if these policies fail to produce measurable results in student outcomes, then a justification for further direct involvement of state control, not to mention the contracting out of educational services to private-sector firms, could result. Privatization and school choice initiatives are policy games in which the payoff for the winners (business and élite interests) usually comes at the expense of the losers (uninformed, undereducated public).

According to Stout, Tallerico and Scribner, educational policy games beg the question: What should be the purposes of schooling? Is the primary purpose of schooling simply to teach the 'three R's'? What should be taught beyond the basics? Given the demonstrated importance of physical and mental health for student success, should we not also try to care for the whole child by coordinating a variety of social services at the school site? Do we have a moral obligation to do so? How will such services be funded? And who will be responsible for the implementation of these services? Given the necessity to bring in local, state, and federal agencies to develop such programs, who will control these initiatives?

In contrast with the privatization game, the movement toward the integration of social services into schools is an educational policy game in which some of the needs of most of the players are actually met. This movement is increasingly viewed as a critical element in the overall improvement of education. As a location containing large numbers of their clients, schools provide a natural clearinghouse for the services offered through public and private organizations. The emphasis on treating the whole child, coupled with the increasing variety of problems affecting today's youth, suggests that multiple services offered through a coordinated delivery system need to replace the current patchwork of support services (Commission on Work, Family, and Citizenship 1988). This educational policy game can offer benefits far beyond the game being played at a given school site.

As one implements collaborative services and interagency partnerships necessary for meeting the needs of most of the players in these games, a significant amount of political conflict may result. One of the most frequent barriers found in the creation of collaborative services is the struggle for power over the nature, direction, and control of such initiatives. Conflicts arising from the development of such partnerships frequently stem from conflicts over competing values and ideologies.

This contention is supported by Fusarelli and Laible (1993) who found conflicts over competing values and ideologies to be one of the principal threats to the creation, implementation, and success of such partnerships. Where the potential for meeting all players' needs exists, conflict can serve to build coalitions of support, mutual accommodation through the exploration of differences, and ultimately to stimulate change. The coordinated social services movement is an example of a positive outcome or positive-sum policy game which some researchers believe produces many more winners than losers.

Policy arenas and policy games: playing the reform game

Playing-field boundaries and political arenas shift from state and federal levels to local school districts and local school sites where educational politics and policy implementation impact on the student. Federal and state activism are examples of the shifting boundaries and arenas within which the game is played. The increased federal and state activism and involvement in education at the local level open up a variety of avenues for macro-level research on the complex interplay of these actors in shaping education policy. A growing tension exists between these institutions as they seek to determine the shape of education in the 21st century. Odden (1991) notes that, on the national level, the federal government is playing an ever-greater role in shaping educational policy.

President Clinton's *Goals 2000: Educate America Act: A Strategy for Reinventing Our Schools* (US Department of Education 1993), calling for the creation of a National Education Standards and Improvement Council, is an example of the increasing involvement of the federal government in shaping the local restructuring of schools. According to Echternacht (1980), this standardization works against the aims of local accountability and autonomy. National standards moderate and limit the autonomy of local interests (Resnick 1981). Therefore, despite the lack of additional funding earmarked for education, it is likely that the federal government, with the cooperation of state governments, will become increasingly involved in influencing the structure and regulation of education.

The movement toward a national curriculum with performance standards to ensure accountability, as conceptualized in President Clinton's Goals 2000 program, provides new areas for research into the federal role in educational policy. Concerns about low student performance, coupled with ever-rising school district tax rates and the increasing unwillingness of taxpayers to support continued bond issues, have renewed interest in increasing educational productivity in schools. Political conflict produced as a result of the increasing emphasis on accountability is sure to increase in the decades ahead. Serious scholarly attention must be given to the effects of this movement on relationships at both the macro and micro level in education.

Of particular concern for many educators is the increasing federal and state emphasis on accountability, in which high stakes rewards are provided for high performance and there are significant consequences for failure. This rhetoric is similar to that of former President Bush in his 1990 State of the Union address in which he emphasized the need to focus on achievement and accountability. The degree of consistency and continuity between the rhetoric of former President Bush and President Clinton's education programs suggests that the federal government is likely to become increasingly involved in shaping local school policy well into the 21st century.

On the state level, Mazzoni notes a trend of state leaders exercising ever more of their plenary responsibilities and authority over the conduct of public education at the local level. High-stakes testing demonstrated to states that they could require a heretofore locally controlled entity (a local public school district) to participate in an accountability context (Baker and Stites 1990). This new era of activated state government in education is likely to create further friction between the interests of local educators and state education officials.

The preeminent question for researchers is how these intergovernmental conflicts are addressed, particularly in light of changing demographic patterns. How, for example, do the bottom-up strategies of decentralization, teacher empowerment, site-based management/shared decision making, and school restructuring square with top-down quality control (Baker and Stites 1990)? These questions are a core concern for researchers studying the politics of education but have not been adequately addressed by researchers in the field. Since these games are not value neutral, intense conflicts erupt over who will control these initiatives and whose vision will prevail; in essence, who will win and who will lose the policy game.

In summary, the decade of the 1980s brought about a renewed interest among state and federal policy makers in educational reform. In fact, the 1980s will be remembered as the decade of reform in education. Buzzwords like outcomes-based education, performance management, site-based management/shared decision making, restructuring, and decentralization were invoked like some holy mantra used to cure a seriously ill patient. It will be up to researchers in the politics of education to determine whether any of these reforms were successful. Preliminary data, however, suggest that reform efforts are spotty. A recent study of high school reform efforts found that only 18 out of 3380 schools are currently implementing all of the following reforms: interdisciplinary teaching, standards and outcome-based education, site-based management, and block scheduling (*Teacher* 1994).

Cuban (1990) notes that few reforms make it past the classroom door. The peripheral problems of schools and the traditional methods of teaching persist despite decades of efforts to alter them. Changing schools is like punching a pillow; they absorb innovative thrusts and soon resume their original shape (Boyd 1987). Given the emphasis among policy makers and educational researchers on comprehensive restructuring of the public schools, these findings do not bode well for the prospects of systemic reform of the educational system.

Moreover, existing tools of understanding are no more than inadequate metaphors for understanding the failure of reforms to alter the regularities of schooling substantially. It is the task of researchers in the politics of education to develop new approaches, methodologies, and theories to explain the persistent failure of school reform. Researchers must gather data on specific reforms and trace their life history in particular classrooms, schools, districts, and regions. More can be done, Cuban (1990) concludes, by studying reforms in governance, school structures, curricula, and instruction over time to determine whether any patterns exist.

Ways to study politics of education games

Throughout the chapters in this yearbook, several authors point out that our field lacks conceptual clarity and that a multiisciplinary perspective is apparent in the way we study the politics of education. Wong suggests that the field has benefited from our multi-disciplinary perspectives. Yet Cibulka indicates that this multidisciplinary perspective weakens the field and urges scholars to pursue work that is both theoretical and applied in nature. In the same vein, Boyd, Crowson and van Geel propose game theory, which is framed within the field of organizational economics, as a vehicle to build the theoretical base in the field of educational politics and policy. Marshall and Anderson offer feminist theories within the cultural studies framework to study the politics of race, gender, and class in schools and educational policy. They indicate that the feminist perspective offers a new lens through which an understanding of the playing fields of education can be viewed.

This section does not pretend to offer a comprehensive coverage of each of the perspectives and methodologies; yet its purpose is to provide avenues for scholars to pursue as lines of inquiry in their future studies. After all, this chapter's intention is to tease out our curiosity and to provide different notions of micro- and macropolitical theory as it relates to the politics of education. It seeks to provide avenues for exploring who plays the game, why and how it is played (the rules of the game), and who wins and loses as a result of the contest. The first perspectives we address is that of political sociology.

Political sociology

Much of the contemporary analysis of education reform ignores the history of education and takes as its rhetoric the definition of change. Scholars and policy makers have initiated education reform assuming that intervention is progress, and that a better world results from new programs, technologies, and organizations which increase efficiency, economy, and effectiveness. The literature is full of instances where programs are identified that seem to be 'successful' as evidenced by principals or teachers (Comer 1980, Levin 1989, and Slavin et al. 1994 among others). Also, policy makers and scholars give value to the perceptions and behaviors of people involved in schools, assuming that the reasons, intent, and practices of those involved in the reform determine the objective outcomes of the reform and change. Yet there is a fundamental element missing in this analysis of education reform: the perspective of political sociology which analyzes education reform as a social and political practice.

Within political sociology, there are several issues or constructs that may be useful for the analysis of education. First, the idea of social regulation is advanced within this perspective. It has been articulated that education reform may be conceived as part of a process of social regulation. For example, Popkewitz (1991) indicates that 'reform is a word whose meanings change as it is positioned within the transformations that have occurred in teaching, teacher education, education sciences, and curriculum theory since the late 19th century' (p. 2). He asserts that reform has no essential meaning or definition and that it does not signify progress in the absolute sense. Popkewitz, however, suggests that reform does entail a consideration of social and power relations. In other words, reform is conceived as social regulation.

Social regulation is a concept not yet fully understood in the politics of education. Yet this concept is critical since schools are social regulatory institutions that construct realities and knowledge for others. Also, schools are social institutions heavily regulated by the state. For example, several researchers indicate that in the last two decades state regulation of the curriculum has increased in many states within the USA (Fuhrman and Malen 1991, Wise 1979).

Thus, education reform has been used as part of a process of social regulation. The tasks of socializing students for a particular purpose (e.g., to get a job at IBM) and the state's form of knowledge (tests) used in schools to shape the views and realities of students are just two examples of the concept of social regulation. Students of social regulation in education may consider asking the following questions: What constitutes reform? What are its changing meanings over time? How are these meanings produced?

Another major construct of the political sociology perspective is that of social epistemology. Epistemology provides a context in which to consider the rules and standards by which knowledge about the world is formed, the distinctions and categorizations that organize perceptions, and ways of responding to the world (Rorty 1979). On the other hand, social epistemology 'takes the objects constituted as knowledge of schooling and defines them as elements of institutional practice, historically formed patterns of power relations that provide structure and coherence to everyday life' (Popkewitz 1991: 15). In other words, the researcher takes the 'knowledge of schooling' and considers the relational and social embeddedness of knowledge in the practices and issues of power in schools. Accordingly, the concern with social epistemology is a political as well as conceptual practice.

Students of social epistemology, for example, question the relation of institutional practices and the regimes of truth as they change over time (Foucault 1980). Regimes of truth are the rules and standards by which individuals define what is good and bad; reasonable and unreasonable; rational, irrational, and nonrational (Foucault 1979). Conceptually, social epistemology makes visible the rules by which certain types of phenomena and social relations of schooling come to be the objects of reform, the conditions of power in these constructions, and the continuities and discontinuities that are embedded in their construction.

The idea is to focus on the social epistemology of schooling as part of power relations. To be specific, students of social epistemology question and study the very nature of knowledge and how such knowledge structures different power relations among those involved in schools. The task of inquiry is to understand which particular social actors maintain their positions of dominance and power, as well as the mechanisms which position them in power, and the elements by which such inequities in power may be eliminated.

Political philosophy and inquiry

In the politics of education field, some have been obsessed with establishing an identity. In doing so, scholars have been profoundly influenced by the literature of the philosophy of science, and they have derived a vision of empirical inquiry from philosophical reconstructions of the character of scientific explanation. This influence has not been carefully examined despite the fact that it has affected nearly every methodological premise of the discipline.

The doctrines of logical positivism and logical empiricism have been especially influential in primary theoretical formulations and in works dealing with inquiry (see Easton's theory of political systems [1965]). In our search for identity we have turned either directly or indirectly to those schools in the philosophy of science for authoritative statements regarding such matters as the goals and standards of empirical inquiry, the criteria of adequate explanation, the meaning of concepts, the character of theory and its relation to observation and factual support, standards of objectivity, and the relationship between facts and values.

This theoretical philosophy and approach to inquiry have led a great majority of scholars to consider only one way to generate knowledge at the expense of many other approaches which may yield a different kind of theorizing about the politics of education and policy decision making. Kuhn (1970) indicated that 'the image of science by which we are now possessed has been extracted from textbooks, from philosophical reconstructions, and from various other accounts of scientific accomplishments rather than from the history of research activity itself' (p. 77).

Whether or not this assessment is correct, it is clear that we need to analyze the historical development of this field. Scholars need to develop new theories of politics in education – new conceptualizations that go beyond logical positivism and logical empiricism (see Marshall and Anderson [Chaper 10] in this yearbook, also Marshall 1993). We believe that logical positivism and empiricism have made great contributions to the field. Nonetheless, we also believe that for any field to grow intellectually, the field needs to be open to new and different perspectives and methodologies.

For example, contextual notions that may define the political behavior of individuals within schools have not been explored as a possibility in the politics of education arena. Contextual theory defines compositional factors, structural factors, and global factors which affect individual political behavior (Books and Prysby 1991, Marshall *et al.* 1989). For instance, compositional factors may include the level of poverty at the community level and its relationship to political behavior of principals or school board members. On the other hand, structural factors would include the degree of property segregation or any structural property of a community.

Books and Prysby (1991) indicate that such variables may affect the political behavior of any members of such a community; thus, it may be useful for scholars in education to consider such a level of analysis to develop and understand the political behavior of education professionals. Finally, global factors are characteristics attributable to the context itself. An example of global factors is media stimuli.

Psychology and political leadership

If one determines that all educational policy makers, including school administrators, are political leaders freely chosen by a community, then it is a standard principle of democratic theory that all leaders be assessed periodically. These two tenets, coupled with the expectation of responsiveness, underlie political accountability (Mitchell *et al.* 1981). Thus, governing boards, legislative bodies and the citizenry-at-large should take seriously the evaluations of leadership performance of educational policy makers. According to Kellerman (1984), there are a number of institutions and individuals who specialize in providing ways to assess political leadership performance. Yet the criteria to assess such a construct are not well established.

Political leadership in education has become more accepted in recent years (Blase 1991, Marshall and Scribner 1991). Thus, political accountability becomes a viable concept to assess political leadership performance in education. This is an area which has not been explored in the writings or empirical work of scholars in the politics of education. One logical area where we may begin this work is with theory, research, and practice which addresses itself to effective and impaired psychological functioning. In this area, one analyzes models developed by psychology and psychiatry over the last ten years that are relevant to the assessment of psychological functioning in political roles. It would be of interest to identify the latest work in this area and further test the ideas espoused in such models.

Postmodernism is another stream of thinking that has gained ascendancy in the humanities, arts, philosophy, and the social sciences. According to Denzin (1991) disciplinary boundaries are nonexistent. For example, literary studies include sociological questions; while social scientists write fiction and so on (see Geertz 1988, Lather 1991). Postmodernism rejects the notion that any method or theory, discourse or genre, tradition or novelty, has a universal and general claim as the 'right' or the privileged form of authoritative knowledge. In fact, postmodernism suspects all truth claims of masking and serving particular interests in local, cultural, and political struggles. According to Scheurich (1994) postmodernism does not automatically reject conventional methods of knowing and telling as false or archaic. Rather, it opens those standard methods to inquiry and introduces new methods, which are then subject to critique. Yet few scholars in the field have fully embraced postmodernism as a lens to study educational policy and the politics of education (notable exceptions include Marshall and Anderson [Chapter 10 of this yearbook], Lather 1991, and Scheurich 1994).

These are some of the possible lines of inquiry for those interested in studying the politics of education. Different notions of macro- and micro-political theory as they relate to the politics of education have been examined and a number of alternative approaches suggested. In the concluding section of this chapter, we make preparations for the future of the politics of education as we move into the next century.

Preparing for the next century

If the game analogy introduced in the beginning of this chapter to describe the politics of education is accurate, then it is a most unusual game we play. Few games have such frequently changing rules, participants, negotiations, compromises, and levels of play as the politics of education. As Frantzich and Percy (1994) suggest at the beginning of the chapter, the action is nonstop. Interestingly, since, as Stout, Tallerico and Scribner claim, there are no permanent solutions to political conflicts in this field, the game never ends! It is a wonder that anyone would want to play at all. But play they do: in every arena, at every level, and in a multitude of ways, as the authors of this yearbook have shown.

The preceding chapters have offered a variety of approaches that scholars may use to study the politics of education during the coming century. These approaches are as varied as their advocates. Iannaccone and Lutz call for a return to research derived from basic democratic theory in the local arena. Likewise, Mazzoni emphasizes the use of multiple perspectives in understanding the impact of educational reforms in federal and state arenas. Further, Wong notes that research in the politics of education has broken away from its parent discipline of political science by not embracing models of rational choice and economic analysis. Boyd, Crowson and van Geel predict that such models will receive increased attention in the future. Others, like Marshall and Anderson, call for research in the areas of power, gender, social class, and race. They suggest that valuable insights could be gained from cultural studies, feminist, critical, and postmodernist approaches to the study of the politics of education. Townsend and Robinson exhort us to challenge our traditional ways of thinking about school governance and to be open to all sorts of new possibilities.

There is agreement among the authors about the need for more theory-building in the politics of education. Cibulka notes the paucity of theory in the area of education policy. Fowler makes similar remarks about the lack of theory in comparative politics. Malen states that the area of micropolitics is a disparate field with little unifying focus. Sroufe argues that traditional political systems analysis is not helpful for understanding federal education policy. He suggests less emphasis on rationality and more on theory devoted to explaining what occurs 'within the box.' As he remarks, rationality does not describe well how decisions are actually made.

In summary, we have reviewed some of the major theoretical perspectives that may be useful to study further the politics of education. With each perspective, there is a methodology or methodologies that may be useful to continue scholarly work in any given area. As Wong suggests, such research is likely to be multidisciplinary in the future. The multidisciplinary scope of inquiry in politics of education scholarship is indicative of a field that is growing in membership, activities, and policy concerns. Combined with the alternative approaches suggested in previous chapters, scholarships in the politics of education has a bright future; and we look forward to the next quarter century where, as Don Layton so aptly puts it at the outset of this Commemorative Yearbook, exciting and interesting scholarship can be expected in this still very young, yet fast maturing field from second- and third-generation politics of education scholars and researchers.

References

BAKER, E. L. and STITES, R. (1990) Trends in testing in the USA, in *Politics of Education Association Yearbook 1990*, 139–157.

BLASE, J. (ed.) (1991) *The Politics of Life in Schools: Power, Conflict, and Cooperation* (Newbury Park: Sage).

BOOKS, J. W. and PRYSBY, C. L. (1991) *Political Behavior and the Local Context* (New York: Praeger).

BOYD, W. L. (1987) Rhetoric and symbolic politics: President Reagan's school-reform agenda, *Education Week*, 6(25): 28, 21.

BOYD, W. L. and KERCHNER, C. T. (eds) (1988) *The Politics of Excellence and Choice in Education* (Philadelphia: Falmer Press).

BROOKS, H., LIEBMAN, L. and SCHELLING, C. S. (eds) (1984) *Public–Private Partnership: New Opportunities for Meeting Social Needs* (Cambridge, MA: Ballinger).

BURNHAM, W. D. (1970) *Critical Elections and the Mainsprings of American Politics* (New York: Norton).

COMER, J. P. (1980) *School Power: Implications of an intervention project* (New York: Free Press).

COMMISSION ON WORK, FAMILY, AND CITIZENSHIP (1988) *The Forgotten Half: Non-College Youth in America* (Washington, DC: W. W. Grant Foundation).

CUBAN, L. (1990) Reforming again, again, and again, *Educational Researcher*, 19(1): 3–13.

DENZIN, N. K. (1991) *Images of Postmodern Society: Social Theory and Contemporary Cinema* (Newbury Park, CA: Sage).

EASTON, D. (1965) *A Systems Analysis of Political Life* (New York: Wiley).

ECHTERNACHT, G. (1980) Title I evaluation and reporting system: development of evaluation models, in G. Echternacht (ed.) *Measurement Aspects of Title I Evaluation* (New York: Jossey-Bass), 1–16.

FOUCAULT, M. (1979) *Discipline and Punish: The Birth of the Prison* (New York: Vintage).

FOUCAULT, M. (1980) *Power and Knowledge: Selected Interviews and Other Writings by Michael Foucault, 1972–1977* (ed. C. Gordon) (New York: Pantheon).

FRANTZICH, S. E. and PERCY, S. E. (1994) *American Government: The Political Game* (Madison: Brown and Benchmark).

FUHRMAN, S. H. and MALEN, B. (eds) (1991) *The Politics of Curriculum and Testing* (London: Falmer Press).

FUSARELLI, L. D. and LAIBLE, J. C. (1993) Of princes and fiefdoms: an exploration of conflict in interagency partnerships, paper presented at the Annual Conference of the University Council for Educational Administration, Houston, TX.

GEERTZ, C. (1988) *Works and Lives: The Anthropologist as Author* (Stanford, CA: Stanford University Press).

GUTHRIE, J. W. (1987) Exploring the political economy of national education reform, in W. Boyd and C. Kerchner (eds) *The Politics of Excellence and Choice in Education* (London: Falmer Press), 25–47.

IANNACCONE, L. (1967) *Politics in Education* (New York: Center for Applied Research in Education).

KELLERMAN, B. (ed.) (1984) *Leadership: Multidisciplinary Perspectives* (Englewood Cliffs, NJ: Prentice-Hall).

KUHN, T. S. (1970) *The Structure of Scientific Revolutions*, 2nd edn (Chicago: University of Chicago Press).

LATHER, P. (1991) *Getting Smart: Feminist Research and Pedagogy with/in the Postmodern* (New York: Routledge).

LEVIN, H. (1989) Accelerated schools: a new strategy for at-risk students, *Policy Bulletin*, 6(May).

MARSHALL, C. (ed.) (1993) *The New Politics of Race and Gender*, 1992, Yearbook of the Politics of Education Association (Washington, DC: Falmer Press).

MARSHALL, C., MITCHELL, D. E. and WIRT, F. M. (1989) *Culture and Education Policy in the American States* (New York: Falmer Press).

MARSHALL, C. and SCRIBNER, J. D. (1991) 'It's all political': inquiry into the micropolitics of education, *Education and Urban Society*, 23(4): 347–355.

MITCHELL, T. R., GREEN, S. G. and WOOD, R. (1981) An attributional model of leadership and the poor performing subordinate, in B. Staw and L. Cummings (eds) *Research in Organizational Behavior*, Vol. 3 (Greenwich, CT: JAI Press), 197–234.

ODDEN, A. R. (ed.) (1991) *Education Policy Implementation* (Albany: State University of New York Press).

POPKEWITZ, T. S. (1991) *A Political Sociology of Educational Reform* (New York: Teachers College Press).

RESNICK, D. P. (1981) Testing in America: a supportive environment, *Phi Delta Kappan*, 62(9): 625–628.

RORTY, R. (1979) *Philosophy and the Mirror of Nature* (Princeton, NJ: Princeton University Press).

SCHEURICH, J. J. (1994) Policy archaeology: a new policy studies methodology, *Journal of Education Policy*, 9(4): 297–316.

SCHLESINGER, A. M. (1986) *The Cycles of American History* (Boston: Houghton Mifflin).

SLAVIN, R. E., MADDEN, N. A., DOLAN, L. J., WASIK, B. A., ROSS, S. M. and SMITH, L. J. (1994) 'Whenever and wherever we choose': the replication of 'Success for All', *Phi Delta Kappan*, 75(8): 639–647.

Teacher (March, 1994) High school reform efforts are spotty (Washington, DC: Editorial Projects in Education), 6.

US DEPARTMENT OF EDUCATION (1993) *The Goals 2000: Educate America Act: A Strategy for Reinventing our Schools* (Washington, DC: US Department of Education).

WISE, A. E. (1979) *Legislated Learning: The Bureaucratization of the American Classroom* (Berkeley: University of California Press).

WOLIN, S. S. (1989) *The Presence of the Past: Essays on the State and the Constitution* (Baltimore: Johns Hopkins University Press).

Index